"Insightful . . . in the footsteps of biologist James Watson and his *The Double Helix*, writing this occasionally gossipy insider's account of the petty jealousies, behind-the-scenes politics, and blatant biases that can play just as important a role in how discoveries are made as the scientific method itself . . . Magueijo also turns out to be a surprisingly engaging and occasionally even artful teacher . . . the insight into what goes on behind closed laboratory doors proves as interesting and educational as the science itself. Magueijo doesn't shy away from saying what he thinks."
—*The Baltimore Sun*

"This provocative new book is upsetting so many pointy-heads . . . As a writer of clear prose, Magueijo is a fine scientist . . . [It holds] mind-bending rewards for readers whose sense of mystery and wonder extends beyond why Trista chose Ryan."
—*Entertainment Weekly*

"A popular account of how VSL theory emerged and what it means. The result is bound to ruffle more feathers; it may actually pluck some. . . . He lucidly explains some very complicated physics, an achievement unto itself. But what makes this book readable is Magueijo's upfront personality and sense of self . . . neither a bookworm nor a wallflower . . . He suffers lesser intellects—real or perceived—poorly . . . his accounting comes across at times as outrageously opinionated and immature, it's also enormously entertaining." —*The San Diego Union-Tribune*

"A memoir replete with political wrangling, name-calling, and academia trashing. He's funny and arrogant (many, many scientists, we learn, are 'senile'), and he offers an insider's view into the strange world of career science. But we get a glimpse too of the romantic side: brilliant people standing around chalkboards, conceiving, then rejecting idea after idea. Along the way we get a spotless primer on some of the coolest ideas in contemporary physics: dark matter, string theory, cosmic rays, quantum gravity, M-theory and its 11 (!) dimensions. If you like to try to fit your mind around some of the most beautiful and complex conjectures about the origins of time, then his book won't disappoint. But just when he has your brain feeling like a Salvador Dali painting, Magueijo will bring you back to earth, putting you in a rave in a tropical India, or accusing a *Nature* editor of penis envy . . . we remember how important it is to question everything, especially our most basic assumptions . . . the more the universe gives up its mysteries, the more outrageously beautiful it becomes." —*The Boston Globe*

"João Magueijo . . . has the markings of a champ . . . an enlightening account of a theory-in-the-making is blackened again and again by a bristly protagonist."
—*The New York Times Book Review*

"[Magueijo's] theory questions the whole premise that the speed of light, the fastest thing in the universe, has always and everywhere been the same. That the speed of light is immutably constant is the bedrock assumption of present day physics . . . if it turns out to be right, it will rock the world of physics. Even if Magueijo doesn't beat the odds, he will have reaffirmed that there's a place in science for irrational pride and bloody-minded stubbornness."
—*Newsweek*

"Magueijo admits that his speculations lack experimental confirmation. So does string theory. But his theory is an attempt to resolve certain mysteries within the Big Bang theory. . . . Magueijo does not suffer fools and bureaucrats gladly . . . crisply written and lucidly argued . . . determined to invest science with new ideas and startling revelations. . . . You can feel the rush of his own excitement and delight. His is a sumptuously unconventional ride into the maelstrom of contemporary cosmology, and it's terrific."
　　　　　　　　　　　　　　　　　　　—The Providence Journal

"Sorry old fogies, but it looks like young radical thinkers are taking over your vaunted institutions. João Magueijo is a thirty-five-year-old anarchist cosmologist physicist . . . If not for the fact that he's a brilliant scientist who developed a revolutionary theory about the behavior of the universe, he'd be just one of the guys."
　　　　　　　　　　　　　　　　　　　—The Portland Mercury

"A gripping, no-holds-barred account of [Magueijo's] challenge to one of the central tenets of relativity and its implications for our understanding of how the universe works . . . a highly readable account. . . . Better still, he gives an honest and revealing insight into what it's like to carry out scientific research."
　　　　　　　　　　　　　　　　　　　—The Guardian (London)

"Magueijo's ideas have come under heavy fire from some physicists, but no one takes issue with his brilliance. Even before he developed an alternative speed-of-light theory, he was among England's most promising physicists, honored by a succession of prestigious scholarships . . . There are tantalizing clues that Magueijo, too, may be onto something . . . Black holes may still be black, but they aren't holes; travel to the stars may not be an impossible dream. The theory even predicts how the universe might end—and how it might be reborn."
　　　　　　　　　　　　　　　　　　　—Discover magazine

"Magueijo brilliantly recounts what he modestly calls 'the story of a scientific speculation,' lucidly explaining a theory that, while certainly not proven, may effectively clear up mysteries about our universe that have puzzled researchers ever since a constant speed of light was written up in that legendary equation of relativity: $E=mc^2$ . . . While Magueijo is right to question elderly Einstein's wish to make the universe conform to his own system of belief, the greatest accomplishment for VSL may be in finding a horizon for scientific investigation, framing a modern-day mysterium tremendum—and opening a space for faith."
　　　　　　　　　　　　　　　　　　　—San Francisco Examiner

"This iconoclastic professor at the Imperial College in London has the courage, or some may say the audacity to challenge that key component of Einstein's deeply ingrained work. . . . Challenging any popular or long-held theory in science is a risk, and can even end a promising career. That's especially true when the challenge is being made to the work of a great scientist like Albert Einstein . . . Magueijo's quirky book is an effort to bring these new ideas to the general public, along with an illuminating peek at these internecine battles among competing scientists and their own stiff conceptions of what's right and what's crazy. . . . Strikingly candid, the book brings esoteric scientific concepts within the reach of nonscientists. Magueijo proves to be a master at taking complex theories and explaining them in simple terms . . . He also provides a thoroughly engaging, if breezy, portrait of Einstein . . . he shows the human side of building a scientific theory, the side behind the black and white

formulas, where the red-blooded, passionate humans argue to defend their ideas and their life's work . . . *Faster Than the Speed of Light* is part science, part biography, and part adventure, a book that takes us behind the closed doors of the intellectual elite and sheds light on the pettiness and brilliance of speculations and theories that ultimately may change our lives and our beliefs."
—*The Christian Science Monitor*

"A fresh interpretation of the mechanics of the universe; it is a new species of science book. . . . [*Faster Than the Speed of Light*] provides the vicarious thrills that the poetically or mystically minded reader looks for in a book on theoretical physics."
—*The Village Voice*

"As we approach the one hundredth anniversary of his annus mirabilis and a three-year traveling Einstein exhibit brightens museums around the world, another upstart egghead has emerged, not to praise phat Albert, but to bury him. Meet Dr. João Magueijo, a thirty-five-year-old theoretical physicist whose variable speed of light theory (VSL) posits that back in the day, light had a whole lot more giddyap."
—*Details*

"Entertaining . . . he denounces the classism, sexism and xenophobia of Britain's ivory-tower elite." —*Time Out New York*

"[Magueijo's] first paper on the topic has racked up 120 citations in a database of physics publications, enough to qualify it as a 'famous' paper. Researchers are seeking him out to explore new theories and to find proof that light has a varying speed . . . the book's take-no-prisoners approach has irked some scientists."
—*Chronicle of Higher Education*

"An iconoclastic, witty Portuguese physicist dares to challenge Einstein's most sacred cow: that the speed of light remains forever constant."
—*East Bay Express*

"João Magueijo has a crazy idea . . . Is he a genius? A charlatan? Both? . . . function[s] very well as a primer for a complex subject."
—*Technology & Society*

"[Magueijo] presents the idea that light traveled faster in the early universe than it does today. If correct, the varying speed of light (VSL) theory solves some of the most intractable problems in cosmology and could have major implications for the study of physics. . . . Recent investigations have shown that whenever the understanding of physics is pushed to the frontier, VSL has something to say."
—*Physical Sciences Digest*

"The heir apparent to Einstein's kingdom . . . The story is a Homeric epic of false starts, dead-ends, office politics, backstabbing colleagues."
—*Seed*

"For its lucidity and persuasiveness, João Magueijo's book on cosmological thinking stands comparison with Simon Singh's *Fermat's Last Theorem* . . . a hip, raucous, hot-blooded, bilious and altogether bewitching exposé of science from the inside."
—*The London Telegraph*

"Like many of the best popular science books, this is not so much a definitive statement as a thrilling report from the front. There hasn't been a writer about science this bolshy since the young James Watson. . . . Fascinating"
—*Time Out London*

"Magueijo unapologetically grinds an arsenal of axes. Editors of scientific journals, college administrators, bureaucrats—he gleefully skewers them all."
—*The Oregonian* (Portland)

"A pugnacious brio that's by turns fascinating and irritating . . . *Faster Than the Speed of Light: The Story of a Scientific Speculation*, a book that reads, at times, as if it were *Jackass Physics: the X-treme Theory of Light* . . . breezy, rude, brash and alarmingly frank, in the manner of someone you meet over pitchers of beer and with whom you're soon spilling not only Bud down your shoes but also your guts in late-night revelations sure to embarrass everyone the next day . . . give[s] the feel of a contact sport to the everyday humdrum of workaday science . . . an odd combination of *Weird Science* and petulant self-centeredness; intelligent, finely drawn scientific speculation and a sophomoric desire to shock—one chapter is titled, 'God on Amphetamines' . . . goes a bit further than being simply Hawking with four-letter words."
—*Inside Denver*

"Educational, provocative, nasty and fun."
—*Mercury News*

"[H]is science is lucidly rendered, and even his penchant for sturm and drang sheds light on the tensions felt by scientists incubating new ideas. This book shows how science is done—and so easily can be undone."
—*Publishers Weekly*

"Magueijo is no crackpot, although many of his colleagues thought so when he suggested a variable speed of light (VSL) . . . VSL verges on heresy . . . the work reads like a battle report . . . Magueijo's obvious lack of interest in pretending to be polite to those he has identified as enemies makes this one of the more scathing scientific memoirs of recent years. . . . The scientific status of VSL remains uncertain, but its creator's account of his investigations is irresistible."
—*Kirkus Reviews* (starred)

"As brash as he is brilliant . . . an often harrowing, frequently absurd intellectual, emotional, and professional adventure . . . exhilaratingly frank in his condemnation of the creativity-killing politics and bureaucracy of science, thrillingly lucid in his explication of cosmology past and present and his own paradigm-altering theory, and utterly compelling in his quest for knowledge and truth."
—*Booklist*

PENGUIN BOOKS

FASTER THAN THE SPEED OF LIGHT

João Magueijo (zh-'wow ma-'gay-zhoo) is a professor of theoretical physics at Imperial College, London, where he was for three years a Royal Society Research Fellow. He has been a visiting scientist at the University of California at Berkeley and Princeton University, and received his doctorate in Theoretical Physics at Cambridge University.

FASTER THAN THE SPEED OF LIGHT

THE STORY OF A SCIENTIFIC SPECULATION

JOÃO MAGUEIJO

PENGUIN BOOKS

PENGUIN BOOKS

Published by the Penguin Group
Penguin Group (USA) Inc., 375 Hudson Street, New York, New York 10014, U.S.A.
Penguin Books Ltd, 80 Strand, London WC2R 0RL, England
Penguin Books Australia Ltd, 250 Camberwell Road, Camberwell, Victoria 3124, Australia
Penguin Books Canada Ltd, 10 Alcorn Avenue, Toronto, Ontario, Canada M4V 3B2
Penguin Books India (P) Ltd, 11 Community Centre, Panchsheel Park, New Delhi – 110 017, India
Penguin Books (N.Z.) Ltd, Cnr Rosedale and Airborne Roads, Albany, Auckland, New Zealand
Penguin Books (South Africa) (Pty) Ltd, 24 Sturdee Avenue,
   Rosebank, Johannesburg 2196, South Africa

Penguin Books Ltd, Registered Offices:
80 Strand, London WC2R 0RL, England

First published in the United States of America by Perseus Publishing,
a member of the Perseus Books Group 2003
Published in Penguin Books 2004

10 9 8 7 6 5 4 3 2 1

Copyright © João Magueijo, 2003
All rights reserved

ISBN 0-7382-0525-7 (hc.)
ISBN 0 14 20.0361 1 (pbk.)
CIP data available

Printed in the United States of America
Set in Garamond
Designed by Lovedog Studio

CONTENTS

1   VERY SILLY   1

PART I   THE STORY OF C

2   EINSTEIN'S BOVINE DREAMS   13

3   MATTERS OF GRAVITY   41

4   HIS BIGGEST ERROR   65

5   THE SPHINX UNIVERSE   79

6   GOD ON AMPHETAMINE   109

PART II   LIGHT YEARS

7   ON A DAMP WINTER MORNING   129

8   GOAN NIGHTS   141

9   MIDDLE AGE CRISIS   165

10   THE GUTENBERG BATTLE   183

11   THE MORNING AFTER   209

12   ALTITUDE SICKNESS   233

EPILOGUE   FASTER THAN LIGHT   259

ACKNOWLEDGMENTS   265

CREDITS   267

INDEX   269

1   VERY SILLY

I AM BY PROFESSION a theoretical physicist. By every definition I am a fully credentialed scholar—graduate work and Ph.D. at Cambridge, followed by a very prestigious research fellowship at St. John's College, Cambridge (Paul Dirac and Abdus Salam formerly held this fellowship), then a Royal Society research fellow. Now I'm a lecturer (the equivalent of a tenured professor in the United States) at Imperial College.

I mention this straight away not because I want to brag but because this book is about an extraordinarily controversial scientific speculation. Very few things in science are as rock solid as Einstein's theory of relativity. Yet my idea challenges nothing less—to extremes that could be perceived as a physicist's career suicide. Unsurprisingly, a well-known popular science tabloid used the title "Heresy" for an article about this work.

From the way the term *speculation* is so frequently used to dismiss ideas with which one disagrees, one might be led to believe that speculation has no role in science. In fact, the opposite is true. In theoretical physics, especially in cosmology, the branch in which I work, my colleagues and I spend a good part of each day trying to punch holes in existing theories and considering speculative new theories that may as well or better accommodate empirical data. We are paid to doubt everything that has been proposed before, to offer crazy alternatives, and to argue endlessly with each other.

I was introduced to this tradition when I became a graduate student at Cambridge in 1990. I soon realized that as a theoretical scientist you spend most of your time interacting with your peers: In a sense, your colleagues take the place of experiments. At Cambridge, semi-informal weekly meetings were convened, where we just argued about whatever had been occupying our minds. There were also the so-called U.K. itinerant cosmology meetings, where at the time, people from Cambridge, London, and Sussex got together to discuss projects that were driving them mad. More mundanely, there was the informal environment of my office, shared with five other people permanently disagreeing and constantly shouting at one another.

Sometimes these sessions would just be general discussions, perhaps focusing on a recent paper someone had just put out. Other times we would go around the room and, rather than talk about new ideas derived from experiments, mathematical calculations, or computer simulations, we would speculate. That is, we would discuss ideas based on no prior experimental or mathematical work, ideas that simply played out in our heads based on a broad knowledge of theoretical physics.

It is a lot of fun to do this, particularly when, after arguing and arguing and finally convincing those around you that you are right, you suddenly slap your forehead and realize that some embarrassingly simple flaw mars your speculation, and that you have just been stupidly misleading everyone for the past hour—or vice versa: You have been childishly taken in by someone else's flawed speculation.

This argumentative tradition puts a lot of pressure on a new graduate student. It can be intimidating, especially when it becomes apparent in the middle of an argument that someone is much more skilled at it than you, and that you are hopelessly out of your depth. And Cambridge, within its ranks of permanent staff, had no shortage of very clever people who loved to show off—people who wouldn't just prove that you were wrong, but who would also let you know that the point you had missed was indeed rather trivial, and that any average Cambridge first-year undergraduate would easily have spotted the error. While these experiences unnerved me, they never depressed me. On the contrary, I found them motivating. You come to feel that unless you think up something truly new you have not earned your place in the community.

During these meetings, one of the topics that frequently came up for discussion was "inflation." Inflation is one of the most popular ideas in current cosmology, the branch of physics that endeavors to answer such profound questions as, Where did the universe come from? How did matter arise? How will the world end? These questions were for a long time a matter for religion, myth, or philosophy. Nowadays they have found a scientific answer in the form of the Big Bang theory, which posits an expanding universe born from a massive explosion.

Inflation is a theory that was first proposed by Alan Guth, a distinguished MIT physicist, and then further refined by several other scientists in response to what we theoretical physicists refer to as "the cosmological problems." Specifically, although virtually every cosmologist now accepts the idea that the cosmos began with a "big bang," there remain aspects of the universe that cannot be explained by the Big Bang theory as we currently understand it. Briefly these problems have to do with the fact that the Big Bang model is unstable. It can only exist as we see it today if its initial state, at the moment of the Bang, is very carefully contrived. Tiny deviations from the magic point of departure rapidly develop into disaster (such as an early death of the universe), and this very unlikely initial condition has to be "put in by hand," rather than deriving from any concrete and calculable physical process. Cosmologists find this very unsatisfactory.

Inflation, which argues that the baby universe expanded unimaginably faster than it does today (so that its size "inflated"), is currently the best answer to these cosmological problems, and to why the cosmos looks the way it does today. There is reason to believe it might be the correct answer; however, there is not yet experimental proof for inflation. And by the most rigorous scientific standards, this means that inflation is still a speculation.

While this does not stop most scientists from being enthusiastic about it, the British theoretical physics community never quite came to believe that the theory of inflation was *the* answer. Call it chauvinism (the theory was first advanced by an American), call it stubbornness, call it science, but if you sat around a table at one of these meetings, inevitably the topic of inflation would come up, and dominating the discussion would be the belief that inflation as we understood it did not resolve certain critical cosmological problems.

Initially, I didn't give all that much thought to inflation because my expertise was in a rather different area: topological defects, an explanation for the origin of galaxies and other structures of the universe. (Defects compete with the inflationary explanation for these structures, yet sadly cannot explain the cosmological problems.) But after hearing again and again that inflation had absolutely no grounding in the particle physics we know and that inflation was merely an American PR success—human nature being what it is—I, too, started to think of alternative explanations.

For the nonexpert it may not be clear why inflation would solve the cosmological problems. Even less obvious is why it is so difficult to solve them without inflation. But to a trained cosmologist, the great difficulty was there, and infuriatingly so, to the extent that no one had succeeded in finding an alternative theory. Inflation had won by default. And for many years, in the back of my mind, and at times even in the front, I puzzled over whether there might be another way, *any* other way, to solve the cosmological problems.

I was into my second year as a fellow of St. John's College (and the sixth of my stay in Cambridge), when one day the answer seemed to drop from the sky. It was a miserable rainy morning—typical English weather—and I was walking across the college's sports fields, nursing a bad hangover, when I suddenly realized that if you were to break one simple rule of the game, albeit a sacred one, you could solve these problems without inflation. The idea was beautifully simple, simpler than inflation, but immediately I felt uneasy about offering it as an explanation. It involved something that for a trained scientist approaches madness. It challenged perhaps the most fundamental rule of modern physics: that the speed of light is constant.

If there's one thing every schoolboy knows about Einstein and his theory of relativity, it is that the speed of light in vacuum is constant.* No matter what the circumstances, light in vacuum travels at the same speed—a constant that physicists denote by the letter c: 300,000 km per second, or as Americans refer to it, 186,000 miles per second. The speed of light is the very keystone of physics, the seemingly sure foundation upon which every modern cosmological theory is built, the yardstick by which everything in the universe is measured.

In 1887, in one of the most important scientific experiments ever undertaken, the American scientists Albert Michelson and Edward Morley showed that the apparent speed of light was not affected by the motion of the Earth. This experiment was very puzzling for everyone at the time. It contradicted the commonsense notion that speeds always add up. A missile fired from a plane moves faster than one fired from the ground because the plane's speed adds to the missile's speed. If I throw something forward on a moving train, its speed with respect to the platform is the speed of that object plus that of the train. You might think that the same should happen to light: Light flashed from a train should travel faster. However, what the Michelson-Morley experiments showed was that this was not the case: Light always moves stubbornly at the same speed. This means

---

*By shining light through appropriate substances it is possible to slow it down, stop it, or, in some sense, even accelerate it. This does not contradict the basic assumption of the theory of relativity, which concerns the speed of light *in vacuum*.

that if I take a light ray and ask several observers moving with respect to each other to measure the speed of this light ray, they will all agree on the same apparent speed!

Einstein's 1905 special theory of relativity was in part a response to this astonishing result. What Einstein realized was that if c did not change, then something else had to give. That something was the idea of universal and unchanging space and time. This is deeply, maddeningly counterintuitive. In our everyday lives, space and time are perceived as rigid and universal. Instead, Einstein conceived of space and time—space-time—as a thing that could flex and change, expanding and shrinking according to the relative motions of the observer and the thing observed. The only aspect of the universe that didn't change was the speed of light.

And ever since, the constancy of the speed of light has been woven into the very fabric of physics, into the way physics equations are written, even into the notation used. Nowadays, to "vary" the speed of light is not even a swear word: It is simply not present in the vocabulary of physics. Hundreds of experiments have verified this basic tenet, and the theory of relativity has become central to our understanding of how the universe works. And my idea was exactly a "varying speed of light" theory.

Specifically, I began to speculate about the possibility that light traveled faster in the early universe than it does now. To my surprise, I found that this hypothesis appeared to solve at least some of the cosmological problems without inflation. In fact, their solution appeared inevitable in the varying speed of light theory. It was as if the riddles of the Big Bang universe were trying to tell us precisely that light *was* much faster in the early universe, and that at some very fundamental level physics had to be based on a structure richer than the theory of relativity.

The first time I threw my solution to the cosmological problems into discussion, an embarrassed silence followed. I was aware that a lot of work needed to be done before my idea could attract some respect; and that, as it was, my idea would look completely crackpot. But I was very enthusiastic about it. And so when I told it to one of my best friends (a physicist, now a lecturer at Oxford), I was not counting on a reaction of complete apathy. But that's what I got: not even a comment—just silence, and then a cautious "hmm. . . ." No matter how hard I tried, I could not draw him into discussing my new idea in the same way that theorists are always discussing even their wildest speculations.

In the following few months, whenever I told my idea to people around me, the reactions were similar. People would shake their heads, at best say, "Shut up and don't be stupid," at worse just be very British and say noncommitally, "Oh, I don't know anything about that." Over the previous six years I had thrown into discussions more than my fair share of mad ideas. Never had I encountered this kind of reaction. When I started labeling my idea VSL (varying speed of light), someone suggested that it stood for "very silly."

You can't take anything that happens at these meetings personally. In fact, the easiest way to drive yourself crazy in science is to take challenges to your ideas as personal insults, even those that are expressed with contempt or venom, and even when you are absolutely sure that those around you think you are a fool. That's science. Every new idea is gibberish until it survives ruthless challenge. After all, what had motivated my idea was precisely my questioning the validity of inflation.

But no matter how many people thought the idea of a varying speed of light stupid, it continued to command my respect if not yet allegiance. The more I thought about it, the better I liked it. And so I decided to stay with it and see where it led.

For a long while it led me nowhere. It is often true in science that a given project does not take off until the right people get together. Most of modern science is done in collaboration. And what I desperately needed at the time was the right collaborator. On my own, I was just going round and round in circles, getting stuck on the same irrelevant details, so that no consistent whole ever seemed to emerge. The whole thing was driving me insane.

The rest of my research work was going well, though, and a year or so later I was overjoyed to find that I had been awarded a Royal Society fellowship. This fellowship is the most desirable junior research position available in Britain, perhaps anywhere. It gives you funding and security for up to ten years as well as the freedom to do whatever you want and go wherever you want. At this stage, I decided that I had had enough of Cambridge, and that it was time to go somewhere different. I have always loved big cities, so I chose to go to Imperial College, in London, a top university for theoretical physics.

The leading cosmologist at Imperial College then was Andy Albrecht. Although he was one of the creators of inflation, Andy had been wondering for years whether inflation really was the right theory. His seminal paper on inflation was also his first paper ever, written when he was still a graduate student. Andy himself has joked that "the answer to all the problems of the universe cannot possibly lie in your first paper"; therefore he had tried and tried to find an alternative to inflation—and like the rest of us he had always failed miserably. We were soon happily working together on the varying speed of light theory. I had found my collaborator.

What followed were years full of the sort of human drama and intensity I had never realized science could supply. This book, in large part, is the story of that journey, as it unfolded from Princeton to Goa, from Aspen to London. It is the story of how scientists work together in a love-hate relationship that sometimes ends well. Of how this mad idea gradually gained shape and reached the stage of a written article. Of how, after submitting our paper for publication, we had to fight with editors and colleagues who were unconvinced that our work was even worth publishing. It is finally the story of why this idea may turn out not to be so mad after all: of how a deeply theoretical speculation may find more experimental support than other better-accepted theories.

But even if this idea is discredited—always a possibility, if not a likelihood, with any intellectual breakthrough—there are several reasons why this story is still worth telling. First, I want people to understand the scientific process for what it really is—rigorous, competitive, emotional, and argumentative. It is people endlessly debating each other, often shouting their disagreements. I also want the non-scientist to understand that the history of science is littered with speculations that sounded great but ultimately did not demonstrate explanatory power and ended up in the garbage bin of scientific inquiry. The process of trying out new ideas, and then accepting or rejecting them, is what science is all about.

But, more importantly, narrating the VSL story will force me to explain in detail the very ideas that the theory contradicts or bypasses: relativity and inflation. Therefore, somewhat paradoxically, you will see them at their best—I have always felt that the most brilliant textbook ideas are best explained by their negatives. Forcing them to undergo cynical challenge, a counterpart to courtroom cross examination, brings them to life.

For these reasons, I believe that you should still read this book even if in the end VSL does not deliver the goods. However, it is obvious that this story will be far more interesting should the goods indeed be abundantly delivered. Naturally, I cannot guarantee that this will come to pass—even though I feel that it is likely.

However, over the past few years several things have happened which indicate that VSL may one day become as mainstream as inflation or relativity. Foremost is the fact that many other people have started to work on it, and in science it's always the case that "the more the merrier." The number of papers written on VSL is increasing every day, and it has started to be included as part of conference topics. A small community is developing around this idea, which pleases me a great deal.

Furthermore, VSL has also left its "cosmological" cradle and has started to solve other problems. Recent investigations have shown that whenever our understanding of physics is pushed to the frontier, VSL has something to say. And indeed, if VSL is correct, black holes may have very different properties than we thought. Collapsing stars could have a totally different demise and die a rather eccentric death. The intrepid space traveller is generally much better off. All in all, there has been an explosion of theoretical work leading to a wild menagerie of new predicted effects, associated with a varying c, whenever physics is forced to confront extreme conditions. Somewhere in the middle of these predictions lies the hope that VSL will be proved true through experiment.

But something even more dramatic may be about to happen. For the last several decades we have known that our understanding of nature is incomplete. Two types of ideas dominate modern physics: the theory of relativity and quantum theory. Each of them is successful in its own realm, but when theorists try to combine them in a chimerical theory called quantum gravity, disaster strikes. We lack the ultimate unified theory, Einstein's unrealized dream of a single consistent framework of knowledge bringing all known phenomena together.

Now VSL is becoming a player in this quest. Perhaps a varying c is the very ingredient that has been missing for so long. This is somewhat ironic: It could be that to fulfill Einstein's dream we will have to give up the one thing he was "sure" of. If so, VSL could be much more than a scientific speculation. It could deepen our understanding of the workings of the universe in ways I had never imagined.

PART I

THE STORY OF C

2   EINSTEIN'S BOVINE DREAMS

WHEN I WAS ELEVEN, my dad gave me a fascinating book by Albert Einstein and Leopold Infeld called *The Evolution of Physics.* In its opening lines, it compares science to a detective story. Except that the challenge is not *whodunnit;* it is why the world works the way it does.

As in any good mystery, the investigators are often led astray. Time and again they must backtrack, separating false clues from real ones. But finally the day comes when a picture emerges, when enough facts have been collected so that they can start applying that uniquely human tool, the power of deduction, to make sense of it all. With a theory of how the mystery arose, and a little luck, they conjecture that certain facts about the case *must* be true. They then test these facts, and, they hope, solve the mystery.

A few paragraphs into that book, however, the mystery analogy is abruptly abandoned. Scientists, we learn, face a dilemma that crime fighters do not. In the mystery of the universe game, scientists never get to say "Case closed." Whether they like it or not, they are never really dealing with one mystery, but rather with one small piece of a huge, interlocked series of mysteries. More often than not, once they solve one piece of the puzzle, that solution suggests that old solutions to other parts of the puzzle are wrong, or at least require reexamination. The game of science can accurately be described as a never-ending insult to human intelligence.

Despite the "indignity" to which it subjects us, I immediately found physics fascinating. I liked in particular the manner in which the mysteries of the universe are posed: The questions asked are often superficially very simple but in reality extremely deep in meaning; they are also beautifully coated in the abstractions of thought experiments and pure logic.

But it was not until I was well into my own career as a physicist that I realized that most problems in physics are not approached in a coolly rational manner; at least not initially. Before we are scientists we are Homo sapiens, a species that, despite its pompous name, is more often driven by emotion than by reason. We don't always carefully sort out false clues and bad assumptions, nor do we limit ourselves to the most rational techniques of problem solving.

During the early development of a new idea we behave rather more like artists, driven by temperament and matters of taste. In other words, we start off with a hunch, a feeling, even a desire that the world be one way, and then proceed from that presentiment, often sticking with it long after data suggests we may be leading ourselves and those who trust us down a blind alley. What ultimately saves us is that at the end of the day, experiment acts as the ultimate referee, settling all disputes. No matter how strong our hunch is, and how well it is articulated, at some point we will have to prove it with hard, cold facts. Or our hunches, no matter how strongly held, will remain just that.

This is particularly true of that branch of physics known as cosmology—the study of the universe as a whole. Cosmology is not about this star or that galaxy; that study is usually called astronomy. Rather, for cosmologists, galaxies are mere molecules of a rather unusual substance that we call the cosmological fluid. It is the global behavior of this all-encompassing fluid that cosmologists try to understand. Astronomy is about trees; cosmology about the forest.

Needless to say, the field is prime ground for speculation. Its puzzles have led us into an elaborate detective story, full of clues, wrong turns, deductions, and empirical facts. Unavoidably, part of the story also shows scientists relying upon hunches and speculations for far longer than most would care to admit.

Cosmology was for a long time the subject of religion. That it has become a branch of physics is to some extent a surprising achievement. Why should a system as apparently complex as the universe be amenable to scientific scrutiny? The answer may surprise you: The universe is, at least in regard to the forces at work, not that complex. It is far simpler, for instance, than an ecosystem or an animal. It is harder to describe the dynamics of a suspension bridge than those of the universe. This realization opened the doors for cosmology as a scientific discipline.

The big leap was the discovery of the theory of relativity, in conjunction with improvements in astronomical observations. The heroes of that story are Albert Einstein, the American astronomer and lawyer Edwin Hubble, and the Russian physicist and meteorologist Alexander Friedmann. Together they blended the constancy of the speed of light and its amazing implications with a larger mystery—the origins of our universe. And it all started with a dream.

WHEN ALBERT EINSTEIN was a teenager he had a very peculiar dream. For many years after he felt intensely marked by this dream, an obsession that would eventually be transfigured into deep reflections. These reflections were to change dramatically the way in which we understand space and time, and ultimately our perception of the whole physical reality surrounding us; indeed, they were to trigger the most radical revolution in science since Isaac Newton, and to bring into question the very rigidity of the space and time into which our Western culture has been embedded.

This was Einstein's dream:

On a misty spring morning, high up in the mountains, Einstein was walking on a path winding its way along a stream dropping from the snowy summits. It was no longer unpleasantly cold, but it was still crisp as the sun slowly started to break through the mist. The birds sang noisily, their songs emerging above the gushing sounds of the tumultuous waters. Dense forests covered the slopes, broken only here and there by gigantic cliffs.

As the path descended further, the landscape opened slightly, and the dense forest started to give way to ever larger clearings and patches of grassy floors. Soon the hanging valleys came into view, and in the distance Einstein could see a multitude of fields, all bearing the unmistakable marks of civilization. Some of these fields were cultivated and divided by fences configured in more or less regular shapes. In others, Einstein could see cows grazing lazily, scattered throughout the meadows.

The sun was now more confidently penetrating the mist, and as it did so it diluted the atmosphere into a tenuous soft focus through which Einstein started to make out details in the fields below. In these parts, it was common to divide the properties with electrified wire fences. They were quite ugly indeed. Also, most of them did not seem to be working at all. Look at all those cows happily munching away the hitherto inaccessible grass on the other side of the fence, their heads pushed through the wires in shocking disregard for private property. . . .

As Einstein reached the nearest meadow, he went to examine the electrified fence. He touched it and, as he expected, felt no shock— no wonder the cows along the fence didn't care about it. As he was playing with the fence, Einstein saw a large figure walking along the opposite side of the field. It was a farmer carrying a new battery and moving towards a shed across the field. Eventually, the farmer reached the shed, and Einstein saw him go inside to replace the expired battery. It was then that through the open door Einstein saw the man connecting the new battery, and *precisely at the same time* he did it, Einstein saw the cows jumping up away from the wire (Figure 2.1). *All at once. Exactly.* A fair amount of displeased mooing ensued.

FIGURE 2.1

Einstein kept on walking, and by the time he reached the farther edge of the field, the farmer was returning home. They greeted each other politely, following which a very strange dialogue took place, the sort of talk one only ever finds in the demented haze of dreams.

"Your cows have extraordinary reflexes," said Einstein. "Just now, I saw you switching on your new battery and, wasting no time, they all jumped up at once."

To this, the farmer looked utterly confused, and stared at Einstein in disbelief. "They all jumped up at once? Thank you for your compliment, but my cows are not in heat. I also looked at them when I switched on the new battery, as I was hoping to scare the living hell out of them: I like to play pranks with my cows. For a short while nothing happened. Then I saw the cow nearest to me jump up, then the next one along, and so on, orderly, until the last one jumped up."

It was Einstein's turn to feel confused. Was the farmer lying (Figure 2.2)? Why should he lie? And yet he was sure of what he had seen: Farmer switches on new battery, first cow jumps up in the air, last cow jumps up in the air, all *exactly at the same time*. Still, no point starting an argument. And for some reason, he was starting to feel like strangling the farmer.

FIGURE 2.2

But then he woke up. What a moronic dream—and with cows, of all animals. . . . And why had he felt pathetically homicidal for nothing? Better forget all this nonsense.

As with many strange dreams, however, some deeper meaning eventually clicks in the dreamer's mind, and indeed, before forgetting the dream altogether, Einstein suddenly had a flash. This was only a dream, and yet, in some sense, it did nothing but exaggerate a real feature of our world. Light travels very fast, but *not* at infinite speed, and what this seemingly innocuous dream was hinting at is that from such a simple physical property of light follows a totally crazy consequence: Time must be relative! What happens "at the same time" for one person may well happen as a sequence of events for someone else.

In fact, light travels so fast that it looks infinitely fast, but that is merely a limitation of our senses. Careful experimentation promptly reveals the truth: Light travels at about 300,000 km per second. The finite speed of sound is more obvious to us because sound's speed is far lower than light's: Sound travels at about 300 meters per second. So, shout at a cliff 300 meters away from you and two seconds later you will hear your echo: Your shout reaches the cliff in one second, is reflected by the cliff, and is returned to you as an echo in another second.

Flash light into a mirror 300,000 km away and two seconds later your "light-echo" will return, a phenomenon well known in radio communications in space, for example in lunar missions. The echo effect on a mission to Mars could be about fifteen minutes: You send a radio message from Earth, it gets there travelling at the speed of light in about fifteen minutes, and the astronauts' reaction gets back to you in another fifteen minutes. Having an argument over the phone while on vacation in Mars is bound to be exasperating.

The cow dream depicts nothing but what actually happens in reality, albeit heavily exaggerated—something that we would indeed perceive with our senses if the speed of light were more like the speed of sound. In Einstein's dream, electricity propagates down the wires at the speed of light.* Hence the image of the farmer switching on the battery travels towards Einstein side-by-side with the electric pulse going down the wire. They reach the first cow simultaneously, and the pulse gives her a shock. Here it is understood that the cow's reaction time is nil,† so that the image of the farmer switching on the battery, the image of the first cow jumping, and the electric signal going down the wire, all now move towards Einstein side-by-side.

When they reach the next cow, she jumps up, and the image of her jumping up joins the parade. Now the image of the farmer switching on the battery, the image of the first two cows jumping up, and the electric signal going down the wire all move towards Einstein side-by-side. And so on, until the last cow. Hence Einstein sees the farmer switching on the battery, and all the cows jumping up, exactly at the same time. Had he put his hand on the wire, he would have gotten an electric shock and said "Scheisse!" precisely at the same time he saw

---

*Artistic license taken here.

†Artistic license taken. Again.

it all happening. Einstein was not hallucinating; it *did* all happen at the same time. That is, at his "own" same time.

The farmer's point of view, however, is rather different. The farmer is subjected to what is really like a series of light-echoes reflected from cliffs/mirrors successively farther and farther away. He switches on the new battery, and that's like a man shouting into an abyss. The electric pulse travels towards the first cow, who jumps up when the pulse reaches her: That's like the shout moving towards a cliff in the abyss and reflecting off it. The image of the cow jumping up, travelling back towards the farmer, is now like an echo returning from the abyss. Thus there is a time delay between his switching on the battery and his seeing the first cow jump; that is, between his shout and its echo. The image of the next cows jumping up in the air is like a series of echoes generated from cliffs farther and farther away, and therefore they have larger and larger time delays—that is, they arrive successively in time.

And so the farmer is not hallucinating either. For him there is indeed a time delay between his switching on the battery and his seeing the first cow jump up. He then sees all the other cows jump up in succession, rather than at the same time. If Einstein had placed his hand on the wire, the farmer would have seen Einstein jumping up and releasing an expletive after all the cows had jumped.

There is no contradiction between the farmer and Einstein, nothing to argue about. The two observers are both saying what they saw, merely reflecting two distinct points of view. If light traveled at infinite speed, Einstein's dream would never have been possible. As things are, it is merely an exaggeration.

And yet, yes, there is a contradiction! Einstein's dream is telling us that there is no absolute concept of "it happened at the same time," absolute in the sense that it must be true for all observers without any ambiguity. Instead, Einstein's dream shows that time must be relative and vary from observer to observer. A set of events that all happen at the same time for one observer may happen as a sequence for another.

But is this an illusion? Or is the concept of time really something more complicated than what we are used to? In our everyday experience, if two events happen at the same time, they happen at the same time for everyone. Could this fact be only a rough approximation? Is this what Einstein's dream was trying to tell him? *Could time be relative?*

THE WORLD EINSTEIN was born into was one in which scientists believed in a "clockwork universe." Clocks would tick everywhere at the same rate. Time was believed to be the great constant of the universe. Likewise space was conceived as a rigid and absolute structure. These two entities, absolute space and absolute time, combined to provide the unchangeable framework for the Newtonian perception of the world: the "clockwork universe."

It is a worldview that resonates throughout our culture. The truth of the matter is that we hate being qualitative, particularly when it concerns financial matters. We prefer to define a unit of money and then refer to the value of anything as a precise number of times that unit.

More generally the definition of units permits marrying the quantitative rigor of mathematics (that is, of numbers) to physical reality. The unit supplies a standard amount of a given type of thing; the number converts it into the exact amount we are trying to describe.

Thus the kilogram allows us to be precise about what we mean by seven kilograms of pineapples, and how much that should cost. The framework of our civilization would not exist without the concept of unit allied to the concept of number. No matter how poetic we claim to be, we love and cannot live without quantitative rigor. I have met very, *very* few genuine anarchists during my lifetime—and I have met some truly weird people.

This philosophy of life permeates our conception of space and time. Space is defined by means of a unit of length—the meter, say. I can then report that an elephant stands along a given road 315 meters onwards, and that means the number 315 times a rigid unit, the meter. We can thus be absolutely rigorous about the elephant's location.

If I want to map a given region on the surface of the earth, I introduce a two-fold spatial structure. I define orthogonal directions, say north-south and east-west. I can then specify exactly where something is relative to me with two numbers: the distance along the east-west dimension, and the distance along the north-south direction. Such a framework defines location precisely, and our obsession with knowing exactly where everything is has found perfect expression in the GPS (global positioning system): Any location on earth can now be labeled to absurd precision by a pair of coordinates.

Naturally, all this is a matter of convention. Australian aborigines map their land by songlines. Australia, for them, is not a one-to-one correspondence between points in the land and pairs of numbers, the coordinates of those points. Rather, their land is a set of highly twisted, multiply intersecting lines, along each of which runs a specific song. Each song relates a story that happened along that path, usually a myth involving humanized animals, contorted fables full of emotional meaning.

At once, the songlines create a complex tangle, so that a point cannot be just a unique pair of numbers; rather, it matters not only where you are (according to our conception) but also where you came from, and ultimately the whole of your previous and future path. What for us is a single point may for aborigines spawn an infinite variety of identities, because that point may be part of many different intersecting songlines. Unavoidably, this creates a sense of property and ownership that does not fit into our culture. Individuals inherit songlines, not areas of land. One cannot build a GPS that operates in songline space.

Yet Australia exists regardless. Songlines stress that to a large extent any description of space is a matter of choice and convention. We choose to live in a rigid and exact space made up of a set of locations, the Newtonian (some would say Euclidean) space.

All these considerations apply similarly to time. A clock is just something that changes at a regular rate—something that "ticks." A tick defines a unit of time. And a unit of time allows us to specify, by means of a number, the exact duration of a given event. What we declare to be a "regular" rate of changing is a matter of convention or definition. And yet, as with many conventions, this is not purely gratuitous; it allows us a simple and precise description of the physical reality around us.

So great is our confidence in our ability to time things that, since Newtonian days, the flow of time has been envisaged as uniform and absolute. Uniform by definition, absolute because why should different observers disagree over the timing of a given event?

Yes, why should they? And yet at the time Einstein had his dream, a crisis was in the making. His dream was a premonitory dream: This rigid conception of absolute space and time was about to be shattered.

ONE TEMPESTUOUS EVENING, the same cows featured earlier in Einstein's dream start displaying unambiguous symptoms of madness. For no reason whatsoever they start moving across the meadow close to the speed of light. Perhaps they are afflicted by an unusual strain of mad cow disease triggered by their earlier electrocution.

The farmer, on hearing the ensuing stampede, turns up in the field with a torch, but when the cows hear him approach, they quieten down and assemble near one edge of the field. But as soon as the farmer flashes his torch at the cows, they start moving away from him at very high speed, closer and closer to the speed of light. The farmer wonders whether his cows might be in heat after all.

But the farmer also wonders about something else. He has just flashed light at cows moving away from him at a speed fast approaching the speed of light. Does this mean that, as the cows nearly catch up with the light, they should see the light rays coming to a halt? This would be very odd indeed—imagine light having a rest. Is there such a thing as stationary light?

To answer this penetrating question, the farmer asks Cornelia, one of the brightest cows in the herd, to inform him of what she sees while running side-by-side with the light rays. She says she sees nothing unusual with the light the farmer has just flashed. It's light like any other. Indeed, Cornelia is very obliging, and just to make sure, she takes all the necessary steps towards measuring the speed of light. She uses standard techniques for measuring this speed, availing herself of clocks and rods that she carries with her. She returns a strange result: She finds that it is business as usual—light moves relative to her at 300,000 km/s.

It is the farmer's turn to feel like strangling Cornelia. By now thoroughly convinced that Cornelia comes from an English herd, the farmer decides to ask two other cows to measure the speed of the light coming out of his torch. But by this time disarray has set in, and the lamest cows are moving more slowly than the others. The cows selected by the farmer are moving away from him at 100,000 km/s and 200,000 km/s. To avoid a proliferation of stupid names, let us call them cow A and cow B (see Figure 2.3).

Given that the farmer sees his light moving at 300,000 km/s he expects these more sensible cows to return the following results: The speed of light should be 200,000 km/s for cow A (that's 300,000 minus 100,000), and 100,000 km/s for cow B (that's 300,000 minus 200,000). It's simple algebra after all. And we have all learnt it in school: Speeds just add or subtract (depending on their relative direction). So to obtain the speed of the light ray with respect to each cow one should just subtract the speed of the cow from the speed of light, right? Or have those cantankerous physics teachers in school been deceiving us all along, as we have always suspected?

Unfortunately, according to our standard perception of space and time, those physics teachers should be right. Let two cars move away from the same location, along the same straight road, at 100 and 200

FIGURE 2.3

km per hour. This means that while my clock ticks away one hour, one car travels 100 km, the other travels 200 km. What is the speed of the faster car with respect to the slower one?

Well, after one hour clearly the faster car is 100 km ahead of the slower one; that's 200 minus 100. So the speed of the fast car with respect to the slower one is 100 km per hour. It's logical enough: You subtract the distances, the time is the same, so you subtract the speeds. What could be controversial about this?

By the same token, if I flash light moving at 300,000 km/s at cows moving away from me at 100,000 km/s and 200,000 km/s, these cows should see my light moving at 200,000 km/s and 100,000 km/s, respectively.

But the cows yet again return a strange result. They both believe that they measure the speed of light relative to them as 300,000 km/s! Hence not only do they contradict the farmer's logic but they seem to contradict each other.

Should we believe the cows? Or should we believe that physics teacher? The good news is that experiment forces us to believe the cows! But that puts us face-to-face with a conundrum. What has gone wrong with our argument showing that the speeds should simply be subtracted? As things stand, what the cows did observe is totally nonsensical.

THIS STATE OF AFFAIRS was more or less the puzzle confronting scientists at the end of the nineteenth century. The experiments supporting the evidence supplied by the cows are now known as the Michelson-Morley experiments. They established empirically the constancy of the relative speed of light, regardless of the state of motion of the observer. If I walk on a train, my speed with respect to the platform has the train's speed added to it. Michelson and Morley found that light flashed from the moving earth still traveled at the same speed: In some funny sense, $1+1=1$ in units of the speed of light. These experiments left physics with a deeply illogical experimental result, one that contradicted the *obvious* and logical dogma that speeds should always be added or subtracted.

This conundrum was solved by Einstein's special theory of relativity. Strangely enough, when Einstein proposed this theory, he was not aware of the Michelson-Morley results. He probably owes more to his dream cows than to these experiments. We will therefore discuss Einstein's solution to this puzzle with reference to his cows.

Let us again employ the services of Cornelia and ask her to stand side-by-side with the farmer. As the farmer flashes light into the fields, Cornelia sets off in hot pursuit at 200,000 km/s. The farmer sees his light ray move at 300,000 km/s. Therefore, in one second he sees light travel 300,000 km away from him, and Cornelia travel 200,000 km away from him. He then *deduces* that Cornelia now sees the light ray 100,000 km ahead of her, and since one second has elapsed, he thinks that Cornelia should see the light ray moving at 100,000 km/s (see Figure 2.4). But when Cornelia is asked to measure the speed of light, she insists that she has found it to be 300,000 km/s. What could have gone wrong?

FIGURE 2.4

It was here that Einstein showed his great genius and courage. He had the audacity to suggest that it could be that time is not the same for everyone. It could be that while one second went by for the farmer, only one third of a second went by for Cornelia. If that happened, Cornelia would have seen the light ray 100,000 km ahead of her, but when she divided that distance by the elapsed time she would have found that the result was indeed 300,000 km/s (see Figure 2.5). In other words, if time goes by more slowly for observers in motion, we can explain why everyone seems to agree on the same speed of light, in blatant contradiction to what is expected from simply subtracting velocities.

But there is also another possibility. Perhaps while one second goes by for the farmer, the same happens for Cornelia, so that time is indeed absolute. Maybe it is space that is playing tricks with us. The farmer sees the light ray 100,000 km ahead of Cornelia because that ray has traveled 300,000 km, whereas Cornelia has traveled only 200,000 km. But what does Cornelia see? It could be that what the farmer perceives as 100,000 km is for Cornelia 300,000 km (see Figure 2.6). If that is so, Cornelia would also measure what she measures: One second has gone by, light is 300,000 km ahead of her according to her rods, therefore its speed with respect to Cornelia, as measured by Cornelia, is indeed 300,000 km/s. But that would imply

FIGURE 2.5

FIGURE 2.6

that moving objects should appear to be compressed along their direction of motion. Could space shrink due to motion?

These are two extreme possibilities, and of course there is a third one: a mixture of the two. It could be both that time passes more slowly for Cornelia, and that her sense of distance is distorted with respect to that of the farmer, the two effects combining to give her the same measurement of the speed of light. While for the farmer one second elapsed and the light ray is 100,000 km ahead of Cornelia, for Cornelia less time has gone by *and* the light ray is farther ahead by Cornelia's rods. In fact, when one works it all out mathematically, one finds that it is indeed a mixture of these two effects that lies behind the dilemma.

This is a crazy way out. But is it true? Sure enough, the farmer soon finds that all this insanity is having an amazing effect upon his cows—they are not getting older! Because time goes by more slowly for fast-moving objects, the farmer gets older and older while his mad cows seem to grow younger by the day. A fast, mad life preserves bovine youth.

He also finds his cows distressingly compressed, nearly flattened into disks when he sees them zipping past. Movement does have a bizarre effect—time goes by more slowly, sizes shrink. Of course, no one has ever tried to measure these effects with cows, but both have been observed with particles called muons, produced when cosmic rays hit Earth's atmosphere.

Clearly something had to give, in the argument leading to the subtraction of velocities. That something was the concept of absolute space and time. Einstein's cows, a.k.a. the Michelson-Morley experiments, shattered the clockwork universe, denying time and space an absolute and constant meaning. Instead, a flexible and relative concept of space and time emerged. The result is cast in what is now known as the special theory of relativity.

WHEN ONE LOOKS at Einstein's solution to the puzzle of light, one is struck by two things: just how cranky it is—and just how beautiful. Who could have come up with such an idea? Who is this guy? One hundred years on we all know who this guy is, but if we rewind the film and look at the way the story played out in 1905, I'm afraid a rather different picture emerges.

Albert Einstein, the young man, was a daydreamer and an individualist. His peformance in school was inconsistent. At times he would do very well, particularly in subjects he liked. At other times disaster would strike; for instance, he failed his university entry examinations the first time around. He hated the German militarism and the authoritarian nature of the education of his day. In 1896, at the age of seventeen, he renounced his German citizenship and for several years was stateless.

In a letter to a friend, the young Einstein described himself rather disparagingly as untidy, aloof, and not very popular. As is often the case with such people, he was perceived as a "lazy dog" by the sensible world (the wording comes from one of his university professors). After he graduated from university he found himself at odds with academia, and a leading professor waged a war against his ever getting a Ph.D. or a university position. Even worse, Einstein found himself at odds with the rest of the world, or, in other words, "deeply unemployed."

At the age of twenty-two, we find him tragically torn. On the one hand he has the cocksure confidence of all free thinkers, and privately lets everyone know just how vacuous he feels the world of respectable attitudes to be. On the other, he has the insecurity of knowing that officially he is a no-hoper, and has to force himself to fawn upon important people in order to get a job. In a letter to a famous scientist, his father paints the following portrait: "My son is profoundly unhappy about his present joblessness, and every day the idea becomes more firmly implanted in him that he is a failure in his career and will not be able to find a way back again."

Despite all efforts, Einstein never made it into academia, at least not until well after he had completed most of the work he is famous for. His early life reads like Jack London's great novel *Martin Eden,* a fact that will forever tarnish the academic world, with its endemic petty games of power and influence. Instead, after many tribulations, a friend and colleague from university days found Einstein a clerk's job at the patent office in Bern, Switzerland. The job was not well paid, but the truth is that there was hardly any work to do.

It was at his patent office desk, at the age of twenty-six, that Einstein flourished—doing little of the work he was supposed to do, while producing, among many other gems, the special theory of relativity.* In a tribute to this university friend, many years later Einstein was to say: "Then at the end of the studies ... I was suddenly abandoned by everyone, facing life not knowing which way to turn. But he stood by me, and through him and his father I came to Haller in the patent office a few years later. In a way, this saved my life; not that I would have died without it, but I would have been intellectually stunted."

"This guy" was therefore someone treading on the margins of society, and, in the end, happy to do so. And who else could have come up with something so apparently insane as the theory of relativity? Unfortunately, in most such cases, especially due to isolation, what comes out are actually cranky and useless ideas. On one of my shelves lie hundreds of letters providing perfect examples of that. At the end of the day we have to give the man credit—he was not just an outsider, he was Albert Einstein. Without him, the world would have been intellectually stunted.[†]

His paper containing special relativity was promptly accepted. The editor of the journal, who made the decision, would later say that he regarded his prompt acceptance of such a loony paper as his greatest contribution to science. But did Einstein realize what he had just done?

---

*Einstein would later suggest that had he gotten the academic post he'd sought, he would never have come up with relativity.

[†]How did Einstein discover special relativity? We know very little about this because he threw away all his rough notes. He did, however, leave one very important clue: He slept close to ten hours a night while performing the crucial calculations. I ascribe to this fact the utmost importance. Popular misconception has it that very intelligent people sleep considerably less than "us ordinary mortals." To support this theory, one usually quotes the noble examples of Napoleon Bonaparte, Winston Churchill, and even Mrs. Thatcher, who could apparently all get away with four hours' sleep. Whether or not these people are prime examples of intelligence I do not wish to discuss here; but I do hope that Einstein's example may forever disprove this pernicious and erroneous theory.

In her old age, Einstein's sister, Maja, would recollect the months that followed in this way:

> The young scholar imagined that his publication in the renowned and much-read journal would draw immediate attention. He expected sharp opposition and the severest criticism. But he was very disappointed. His publication was followed by an icy silence. The next few issues of the journal did not mention his paper at all. The professional circles took an attitude of wait and see. Some time after the appearance of the paper, Albert Einstein received a letter from Berlin. It was sent by the well known Professor Planck, who asked for clarification of some points which were obscure to him. After the long wait this was the first sign that his paper had been read at all. The joy of the young scientist was especially great because the recognition of his activities came from one of the greatest physicists of that time.

IN FACT, WHAT HE HAD just done was far reaching in many different ways, well beyond the introduction of a relative space and time. Relativity went from success to success, and Einstein's early misfortunes soon came to an end as the world recognized his great achievement. The implications of relativity were immense, and as I have already said, to some extent the language of physics is nowadays the language of special relativity. But this is not primarily a book on relativity, so let me just highlight what I consider to be the three most important consequences of the theory.

The first is that this constant speed, the speed of light—which is the same for all observers at all times and in all corners of the universe—is also a cosmic speed limit. This is one of the most perplexing effects predicted by the theory of special relativity, but it results logically enough from its foundation principle. The proof goes as follows: If we cannot accelerate or brake light, we also cannot accelerate anything traveling more slowly than light all the way up to the speed of light. Indeed, such a process would be exactly the reverse of decelerating light, and if it were possible, its mirror image would also have to be possible, in contradiction with special relativity. Thus, the speed of light is the universal speed limit.

This fact may sound strange, but physics is often counterintuitive. Indeed, science fiction films are fond of showing spaceships breaking the speed-of-light barrier. According to relativity, it's not so much that you would get a cosmological speeding ticket; more to the point, relativity shows that you simply would not have enough power to do it, no matter what the nature of your engine.

The existence of a speed limit has a tremendous impact on the way in which we should perceive ourselves in the universe. Our nearest star, Alpha-Centauri, is four light years away from us. This means that no matter what our state of technology, a round trip there will take us at the very least six years, as measured on Earth.* For the astronauts, this could mean only a fraction of a second, due to the time dilation effect. So at the end of the trip there could be a six-year mismatch between the ages of the astronauts and the loved ones left behind. This could perhaps be cause for a few divorces; but nothing more serious, one would hope.

Yet that is our nearest star, barely around the corner in astronomical terms. What about something more considerable on a cosmological scale? Well, let's not even be that adventurous; let's consider a trip just to the other side of our own galaxy. Even that is thousands of light years away from us. Hence, pushing technology to the very limits, a trip there and back would take several thousand years as measured on Earth. We should also make sure that the time dilation effect is such that for the astronauts this corresponds to at most a few years, if we don't want such a space mission to become an ambulatory cemetery.

---

*I am ignoring the significant issue of how to accelerate the astronauts up to and down from close to the speed of light. This has to be done as fast as possible but without killing anyone, and may turn out to be the greatest limitation.

But therein lies the catch. If I push technology to the limits and attempt such a round trip close to the speed of light, in astronaut time it is possible to cover huge distances in a few years, but they will always correspond to thousands of years on Earth. What a pointless space mission! By the time the astronauts get back to Earth, they might as well be visiting another planet. This is no longer just a matter of a few divorces: These poor astronauts would be completely severed from the civilization they had come from.

If we want to avoid such disasters we need to keep well below the speed of light, and so not stray far from home. Our range has to be much smaller than the speed of light times our lifetime—say, some dozens of light years, a ridiculous figure by cosmological standards. Our galaxy is a thousand times bigger than that; our local galaxy cluster is a million times larger.

The overall image to emerge is one in which we are confined to our little corner of the universe. A bit like life on Earth if we could not move faster than a meter per century—awfully limited roaming ability. This is very depressing indeed.

A SECOND IMPORTANT IMPLICATION of the theory of relativity is the conception of the world as a four-dimensional object. Usually, we conceive space as having three dimensions: width, depth, and height. What about duration? Yes, to some extent anything also has a "time-depth," or duration, but we know that time is essentially different from space. So including time in the count or not is essentially an academic matter. Or it *was*, before the theory of relativity.

According to relativity, we find that space and time are observer dependent, that duration and length may dilate and shrink depending on the relative state of motion of the observer and the observed. But if space shrinks when time dilates, isn't it as if space were transforming into time? If that is so, then the world is indeed four-dimensional. We cannot leave time outside the count, simply because space may be transformed into time and vice versa.

Such is the perception of the world nowadays called Minkowski space-time (as in the same Prof. Minkowski who once labeled his student Albert Einstein a lazy dog). Space and time, according to relativity, are no longer absolute; but a mixture of the two—"space-time"—remains absolute. It's a bit like the theorem of conservation of energy, which we all have learnt in school. There are many forms of energy, including, for instance, movement and heat. Each form is not conserved by itself, because we can transform, say, heat into motion (e.g., using a steam engine). However, the energy's grand total is conserved and always remains the same. Likewise, space and time are no longer constant, but depend on whom you are talking to. Depending on the observer, duration and length may be dilated or shrunk. But the grand total, space-time, is the same for everyone.

This space-time picture is quite revolutionary if you think about it a bit. The basic unit of existence is no longer a point in space, but the line depicted by this point in space-time when you consider it at all times—what Minkowski called the *world-line*. Hence think of yourself not as a volume in three-dimensional space, but as a tube in four-dimensional space-time, a tube consisting of your volume shifting forward in time, towards eternity. In a fit of nerdy witticism the physicist George Gamow titled his autobiography *My world-line*.

ONE LAST IMPLICATION of special relativity I wish to highlight is the famous $E=mc^2$ equation: Energy equals mass times the speed of light squared. This is probably the best-known physics formula nowadays. How did it come about?

The derivation is in fact closely related to the proof that the speed of light is a universal speed limit. A few pages ago we proved this fact *logically* (showing that if we could accelerate something up to the speed of light, then conversely we should be able to decelerate light, in contradiction with the constancy of c). This is fair enough, but *dynamically,* what is the reason we cannot overtake light?

If you push something you create an acceleration, that is, a change in the speed of an object. However the larger the object's mass (colloquially, the heavier the object), the larger the force you need to produce the same acceleration. What Einstein found was that the faster an object appears to move, the "heavier" it feels (or, noncolloquially, the larger its mass).* He also found that as a given object is seen to approach the speed of light, its mass is seen as becoming infinitely large. And if the mass of an object becomes infinite, no force in the universe is big enough to produce noticeable acceleration anymore. Nothing can produce that little extra bit of acceleration that would push an object all the way up to, and beyond, the speed of light.

This is why the speed of light works as the cosmological speed limit. You run out of steam as you try to do something illegal. What you are pushing gets heavier and heavier so that you can't push it hard enough to break the speed of light barrier and get that cosmological speeding ticket, whether you want to or not.

And what does this have to do with E=mc²? What follows is Einstein's mind at its purest, guided by disarmingly simple reasons of symmetry and aesthetics. He now notes that motion is a form of energy, sometimes called kinetic energy. If by adding motion to a body you increase its mass, it looks as if by increasing its energy (here in the form of movement) you increase its mass. But what's so special about that energy's being in the form of movement? We know that we can convert any form of energy into any other. Then why not say that by increasing the energy of a body (in whatever form), we increase its mass?

It's a brave generalization, but it has implications that should in principle be observable. Heat an object and its mass should increase. Stretch a rubber band and it accumulates elastic energy; therefore its mass should also increase. Not by much, but a little bit. And so on,

---

*The subtle distinction between weight and mass is behind the formulation of the general theory of relativity, to be described in the next chapter.

for all forms of energy. Thus, in a great coup of insight, Einstein, in a three-page paper published in 1905, proposes that by increasing the energy of a body by E, its mass m should increase—by E divided by the square of the speed of light:

$$m = E/c^2$$

The argument rests on the fact that when kinetic energy is added to a body, its mass is increased, and that for reasons of symmetry this should be valid for all forms of energy.

But then there was a real brain wave, two years later, in 1907. Einstein pushed his sense of beauty and symmetry even further, for the better—or worse—of all of us. Two years before, he had noticed that confining a relation between increases in mass and energy to energy in the form of motion spoiled unity: all energy should increase the mass of a body. But doesn't this seem to imply that energy already has a mass, or, even better, that the two are the same thing?

Identifying any form of energy with mass and vice versa seems to improve the oneness, the perfection of the theory. But then if all forms of energy carry a mass, shouldn't mass also carry energy? Shouldn't mass, in fact, be identified with a form of energy? Thus Einstein did something horribly simple to the formula above. He rewrote it as

$$E = mc^2$$

It looks brutally simple and yet it is a gigantic conceptual leap. Again it is a brave generalization, but not a gratuitous one. It has observational predictions, it can be tested. When you put numbers into this formula and perform a short calculation, the implication is that inside 1 gram of matter lies dormant an energy equivalent to the explosion of about 20,000 kg of TNT.

But that's obviously wrong, isn't it? How did Einstein cope with this enormous contradiction? No worries. Einstein pointed out that we do not notice energy, but only variations in energy. I feel cold if thermal energy moves from my body into the environment. I feel my car accelerating if I push the accelerator pedal and burn fuel, thereby taking chemical energy from the fuel and converting it into motion. The tremendous amount of energy harnessed inside 1 gram of matter passes unnoticed because it is never released into the world; it's just like a huge reservoir of energy sitting inside a body, never making its presence known.

In a popular science account of this concept, written by Einstein himself, he considers by analogy a phenomenally rich man who never parts from his money. He lives modestly, and goes around spending only small sums. Thus no one knows of his large fortune because only variations in his wealth are perceptible to the world. The large energy associated with the mass of objects is very similar.

Perhaps I should remind you that while all this was happening, nuclear physics was hardly a subject. The whole concept of the energy of mass evolved from pencil and paper, and, ironically, from considerations of symmetry and beauty. Little did Einstein the pacifist know what he was about to unleash.

On August 6, 1945, Einstein's "phenomenally rich man" bestowed his grim fortune upon the world.

THE THEORY OF RELATIVITY was a massive intellectual earthquake. Today, no one disputes that relativity revolutionized physics, but it also forever changed our perception of reality, not to mention its dramatic effects on the course of twentieth-century history. So much so that nowadays everyone has heard of Einstein's theory of relativity.

But Einstein was not finished yet. He soon realized that his theory was incomplete, which is why he called it "special" relativity. He therefore set out at once to find the complete, "general" theory of relativity. This turned out to be even more groundbreaking and mindblowing. But the story of its discovery is not so straightforward:

Teenage, dreamlike innocence was lost at this point, and Einstein's struggle for the general theory of relativity would be a very adult nightmare indeed. Looking at photographs of Einstein, taken around the time he finally finished the general theory, we see an utterly exhausted man. He has the look of someone who has just emerged from a lengthy and bloody intellectual battle.

3   MATTERS OF GRAVITY

EVERYONE HAS HEARD OF Einstein's theory of relativity, but not everyone knows that there are actually two theories of relativity: the special and the general. You have just learnt about special relativity; in fact, this theory is valid only in circumstances in which we can ignore the force of gravity. Of course, those are very "special" circumstances. Under more "general" circumstances gravity is important: Think of what holds us on Earth, what dictates the motions of planets, or, more to the point of this book—VSL being a cosmological model—what controls the life of the universe as a whole. Thus the need for a general theory of relativity, valid even in circumstances in which gravity cannot be ignored.

The general theory of relativity turned out to be a very different ballgame from the special theory. In 1905, fresh from proposing special relativity, Einstein already knew that his latest offspring could not be a valid description of nature when gravity was involved. He also knew that the then-accepted theory of gravity, Newton's theory, was inconsistent with special relativity, with the constancy of the speed of light, and with the idea that time was relative. But finding a "relativistic" theory of gravity proved a gigantic struggle, even for the great man.

Sadly, none of the experience amassed in the building of the special theory was of any relevance to the general theory, and it would take Einstein ten years of hard work to arrive at the final product. In 1912 he would say, "I occupy myself exclusively with the problem of gravitation, and now believe that I will overcome all difficulties with the help of a friendly mathematician. . . . Compared with this problem, the original theory of relativity is child's play."

Indeed, it was an ambitious enterprise. It required mathematics well beyond his skills, so much so that he would feel the need to enlist the services of professional mathematicians. He would make mistakes, correct them, then make them again. By chance he would stumble upon the right theory, and naturally enough, abandon it. Finally, he would get back to the right theory. The whole thing reads like a comedy of errors, but with the right ending, an ending only a complete genius could have contrived.

Along the way, in 1911, Einstein even proposed a varying speed of light theory! Nowadays, scientists are either horrified by this paper written by the great Albert Einstein while a professor in Prague, or they simply don't know about it. Banesh Hoffmann, Einstein's colleague and biographer, describing this piece of Einstein's work, puts it this way: "And this means—what? That *the speed of light is not a constant.* That gravitation slows it down. Heresy! And by Einstein himself."

This is a revealing and highly amusing reaction. It seems to me that contradicting textbook wisdom is only heresy for those who have learnt it from the textbook. If you come up with a textbook idea yourself, you are a lot less religious about it. But let me hasten to point out that this 1911 VSL theory has nothing to do with the one I am writing about, proposed at the end of the twentieth century. It was the wrong theory, and Einstein happily threw it into the garbage bin together with quite a few other dead ends.

It wasn't until 1915, during World War I, that Einstein finally reached what we now know as the general theory of relativity. The result is a monument to human intelligence, a cathedral of mathematical ingenuity and powerful physical insight. Without it, modern cosmology (or VSL—or this book) would not exist.

It is also a phenomenally complex theory, requiring the use of a totally new piece of mathematics never seriously used before in physics, *differential geometry*. It is a very difficult theory to understand unless you are a professional physicist. My own somewhat contorted early relations with the intricacies of general relativity are testimony to this fact.

AFTER I READ THAT BOOK by Einstein and Infeld when I was eleven, I decided that I wanted to know about relativity in greater detail; in particular, I wanted to see the equations, not just a pile of words. The chance came along when I found an excellent book by Max Born exposing special relativity using mathematics, but only of the sort one learns early on in school.

That was just the book for me. If you hate mathematics you probably won't understand this, you won't see why anyone would want to learn about something by means of formulae when a description using words is available. But that is the physicist's mind for you, and I was already thinking like a physicist. We don't feel that an idea has properly become a physical theory until we see it cast within the framework of mathematics. As Galileo once said, the book of nature is written in the language of mathematics.

With glee, I carefully followed all the mathematical derivations in Born's book, and when I finished the chapter on special relativity I felt that I had finally sunk my teeth into it. But when the book got into general relativity, it suddenly became very vague and wordy. I felt that I had once again reverted to the unpalatable level of mere words, and lost my grip on the subject.

A proper technical treatment of general relativity may be found in yet another book, *The Meaning of Relativity,* written by Einstein from lectures given at Princeton University in 1921. One day, my best friend turned up at school with a copy of this book. Even though we did not understand a word, we marveled at its complexity: so much complicated mathematics, such impenetrable arguments. . . . Unwisely I thought again that this was just the book for me.

I rushed to the bookshop where my friend had bought his copy, but to my great disappointment the shop assistants refused to sell me their last copy. They also informed me that it was a very rare edition, well out of print. I was quite annoyed at the time, but in hindsight I have to give them credit. On your shelves lie the last two copies of a somewhat rare and very technical book by Einstein, and two kids turn up trying to buy them. . . . To this day I still wonder what they thought we wanted the book for: perhaps for help in building a nuclear bomb. Certainly they felt we were up to no good, which to some extent was true.

But at the time I thought I had just been the victim of blatant discrimination—ageism, if you want to call it that. So I fought back, and asked my dad to go and get the book for me. Initially, he agreed, but the next day came back empty-handed, shaking his head in disapproval. He told me that "it was unsuitable reading for children," which made me wonder whether he had correctly understood which book I had asked him to buy. But he went on to say that Einstein's book would just confuse me because I would not know the meaning of "all those symbols, all those parameters." I raised the proverbial racket, well-known to those with children, until my dad, hoping to get some peace, went to the bookshop and bought it for me.

A lot of hard work followed, but after trying and trying, I found that I still understood exactly nothing of the contents of that book. I felt rather stupid, but it also became obvious to me that the problem was that, unlike Born's book, this one required more than just the kind of mathematics one learns in school. It required advanced calculus, mathematics one usually learns only at university, of which I knew next to nothing. Thus, in this early experience, I came to grips with the fact that the general and special theories of relativity are two very different stories.

However, I refused to give up, and decided that the first step would be to learn calculus by myself. And so I collected various books on the subject, and over the following years studied them in detail. I then developed an unusual ritual: Every six months or so I would reopen Einstein's book to check whether my improving mathematics already allowed me to understand anything in that book, no matter how trivial. And predictably, I would always find myself still completely in the dark.

This early trauma is responsible for most of my mathematical education. I acquired almost all my calculus toolbox by tutoring myself up to the level where I hoped I could understand that book. But as I started to run out of new mathematics to learn, and still couldn't understand a word, I gradually gave up all hope. I went to university, became a physicist, and the book's unopened pages became brittle as I gave up on grasping "the meaning of relativity."

Many years later, already a physicist in Cambridge, I accidentally came across that old copy of Einstein's book, left forgotten at my parents' home. I opened it, and suddenly it all became clear to me. I could not understand a word, not because I did not know the relevant background mathematics and physics, but because the notation used was impenetrable.

Indeed, perhaps as a result of his forced isolation from academia at the start of his career, Einstein used a very baroque set of symbols, which no one else has ever used, either at the time or since. The speed of light, instead of a c, becomes a V. $E=mc^2$? Nah . . . $L=MV^2$ is much better. These two examples may not be so difficult to figure out, but by the time you get to general relativity, the result is like coded text: lines of multiple integrals, gothic letters galore, tensors written out as full matrices—the perfect caricature of the mad scientist's scribblings! To have any hope of understanding any of it you have to start by breaking the code.

Naturally, once I had learnt general relativity by other means, I could recognize it in that book, work backwards, and decipher that crazy notation. But if you had a first crack at general relativity by using that book you did not stand a chance, whatever your background. The book might as well have been written in Chinese.

Thus it turned out that my dad was right all along, although perhaps for the wrong reasons. Indeed I would not know the meaning of "all those symbols, all those parameters." But as often happens you end up conquering Mount Everest while fighting for the Moon. Besides, children never listen to their parents. . . .

IN 1906, EINSTEIN was well aware that the Newtonian theory of gravity was at odds with his theory of special relativity at a very fundamental level. It contradicted the idea that nothing could travel faster than the speed of light. This is not very difficult to understand.

The force of gravity is one of the most obvious in our daily lives; it keeps us from flying up into the sky, for a start. But gravity is different from all other forces in our daily lives in one very important aspect. All other forces seem to be contact forces. If you punch someone, that person has few doubts that contact has been made. And all else, pushing, pulling, friction, and so forth, all the mechanical forces that surround us, seem to take effect by direct contact, so much so that the idea of force as an action resulting from contact pervades our everyday conception of force.

The one apparent exception is gravity, which seems to act at a distance. When I jump from a diving plank there are no ropes attaching me to the Earth, and still the Earth pulls me towards its center. The Sun pulls the Earth around its orbit from 100,000,000 miles away, again without ropes attached. These facts puzzled Newton so much that he vented his frustration about his own theory in the following manner: "That gravity may be . . . so that one body may act upon another at a distance through a vacuum, without the mediation of something else, by and through which their action and force may be conveyed from one to another, is to me so great an absurdity, that I believe no man . . . can ever fall into it." Clearly Newton would have been much happier if the Earth and the Sun were indeed tied together by ropes.

Of course the idea of action at a distance is only superficially puzzling; with a bit of thought you will realize that all actions, even those we associate with contact, are actually actions at a distance. Did you really make contact with that punch? Try to visualize the molecules you are made of. Perhaps think of these molecules as little solar systems, governed by electricity instead of gravity, which repel each other when brought close together. They don't really touch each other; they repel, at a distance, when brought sufficiently close, and that's what produces what feels like contact in that punch. It's even possible that a few molecules were ejected from your knuckles and someone else's face; but there was certainly never any real *contact* between those molecules.

Mechanical contact forces, seen from a molecular level, are therefore also actions at a distance, albeit of an electrical type. And indeed at a fundamental level, all everyday forces are actions at a distance, of either the gravitational or the electromagnetic variety. There are several differences, however, between these two types of forces. Electric forces may be shielded at long range because objects may be electrically neutral; in contrast, nothing is gravitationally neutral. Electric forces are also much stronger than gravity: You have to collect quite a lot of mass before gravity becomes of any significance. A good illustration is the fate of a man jumping out of a plane without a parachute. Gravity takes quite a while to accelerate him, but the electric forces he meets when he crashes into the ground decelerate him very quickly indeed.

However, in 1906 there was another crucial difference. "Electric" types of action were known to travel with the speed of light. Indeed, the whole of special relativity is associated with the electromagnetic theory of light, rather than with cows as I may have led you to believe. Newtonian gravity, on the contrary, was envisaged as an instantaneous action at a distance. And therein lay the contradiction between Newtonian gravity and special relativity. According to special relativity, nothing can propagate faster than light, let alone with infinite speed.

The contradiction is even more fundamental than it looks. In Newtonian gravity, if the Sun changes position, the Earth will know about it, via the force of gravity, straight away, that is, "at the same time." But stop! We know that in special relativity the concept of "same time" is relative and takes different meanings for different observers. Therefore, a theory telling you that a force takes effect at the same time cannot be consistent with special relativity, for we know that action must take an absolute meaning, be the same for all observers, to avoid contradiction.

These were the difficulties presented to Einstein, in the face of gravity on the one hand and his special theory of relativity on the other. He needed to replace Newton's *instantaneous* action at a distance with a theory in which gravity traveled with a finite speed, which, for simplicity, should also be the speed of light. Sound simple? It always does after someone has cracked the Columbus egg. In fact, for various technical reasons, gravity refused to travel with the speed of light, and Einstein was left fumbling in the dark for a long time.

At last inspiration came—from an old experiment attributed to Galileo, which no one had ever completely understood.

ON A FINE CORNER of the Italian town of Pisa stands a monument to the human ability to blunder—a leaning tower, which some wager will not stand for much longer despite all attempts to fix it with modern technology. The tower began to lean right from the beginning, while the first few storeys were still being built. It is not generally known that in those early days the tower actually leant in the opposite direction. In an attempt to deal with the subsiding foundations, however, the engineers became a bit overenthusiastic, and soon enough the tower started to lean the other way—in its present direction.

As more storeys went up, attempts were made to disguise this defect by building the new storeys horizontally once the subsidence was taken into account. As a result, the middle sections of the tower are shaped like a banana. This trick may have worked at first, but as the subsidence has been steadily increasing throughout the centuries, the banana shape of the tower is now painfully obvious.

The leaning tower of Pisa is a comedy of errors, a bit like the years that led Einstein to general relativity. Except that with the tower you are left with the errors, whereas with general relativity, only the final product is remembered.

It was, apocryphally, from the top of this tower that Galileo performed the famous experiment in which he threw down a variety of heavy objects, all equally smooth (so that they shared the same air friction), but of different weights. He found that they all took the same time to fall, traveling down at the same speed. This was in contradiction to Aristotle's physics, which enshrines the "common sense" notion that heavier objects fall faster than lighter objects. But remove friction from all considerations, and indeed heavy and light objects, subject *exclusively* to the force of gravity, fall at exactly the same speed.

Not convinced? Take a sheet of paper, put it on top of a larger book (so that the top of the book contains the whole of the sheet) and drop them. You will find that book and sheet will fall together.*

This strange fact is very counterintuitive and people react against it quite passionately at times. I recall once standing on top of a diving plank with my sister and another guy who wondered what would happen if the plank broke and we all fell. He felt that it would be a total disaster, because the plank, being heavier, would fall ahead of us, and then we would fall on top of it. A great argument broke out until my sister, who was more interested in flirting with the guy, told us to shut up with all that nonsense.

It was this weird phenomenon that provided the inspirational roots of general relativity. First of all because it punched a hole in the very theory that Einstein was trying to replace: the Newtonian theory of gravity, which was never fully able to account for why light and heavy objects fall with the same acceleration. And in science, as in detective stories, before you can find the correct solution to the mys-

---

*This is a very misleading experiment, but failing a trip to the Moon, it serves to make the point. Astronauts on the Moon would see a hammer and a feather fall at the same speed.

tery you first have to find the flaw in the prevailing but wrong theory, the "false solution" that landed the innocent guy in jail and left the real criminal free.

Here's how Newton attempted to explain why everything falls the same way: "Larger" or more "massive" objects resist forces more, as we know. This resistance is called inertia, and is measured by the so-called *inertial mass*. The larger the inertial mass of an object, the larger the force required to give it a certain acceleration.

But gravity counters this effect with a peculiarity: It pulls more massive bodies harder, so that the larger the body the larger the force of gravity. This fact is measured by the weight, or *gravitational mass* of an object. Now it just so happens that the gravitational and inertial masses of all objects are the same, a fact so obvious that one seldom even notices that this need not be true.

Thus the "larger" and "denser" an object is, the greater its inertia (that is, its resistance to being accelerated), but its weight is then also greater and so is the force of gravity acting on it. Therefore, the body resists gravity more, but gravity also pulls it harder, the two effects combining *perfectly* so that the same acceleration is imparted upon all objects regardless of their masses.

Why is this a big hole in the Newtonian theory of gravity? Because the theory provides no explanation for the exact equality of the inertial and gravitational masses. In Newtonian theory, this equality is a coincidence, even a curiosity. By observation, we find between two rather different quantities an exact equality that applies to *all* objects without distinction, yet our theory cannot offer an explanation for this striking fact. It merely states that it is true.

Still, the successes of the Newtonian theory of gravity were and *are* so great that for many centuries no one really cared about this conceptual deficiency. At some point, a major factor in the success of every theory is whether it is operationally correct. And to this day, rocket launches are based on Newton's theory of gravity, and no one has ever gotten lost in space.

THIS VIEW WAS NOT SHARED by Einstein, who was quick to realize they had jailed the wrong guy as he drew attention to this conceptual hole in Newtonian theory. He started to wonder: Does the fact that all bodies fall the same way mean something?

I know this may sound demented, but let's try to subject ourselves to the following vision. Let's think of all the objects that are governed by nothing but gravity—the planets around the Sun, comets flying through the solar system, rocks falling from the skies . . . and now go totally mad and imagine that the whole of space and time, space-time, is actually filled with imaginary free-falling objects. Each point in space-time has its own free-falling creature, one for each direction and all possible speeds. As we have just found out, it does not matter which beast is allocated to which point because they all fall the same way, and all follow a trajectory that does not care about its passenger. So much so that it seems as though the lines depicted by this swarm of free-falling creatures don't depend on who or what is falling but are a property of the space-time in which they are falling, a space-time pervaded by gravity.

These lines are in general rather curved because a basic property of gravity is that it pulls objects' trajectories away from straight and uniform motion. And now get ready for the great conceptual leap, the swell of yet another massive brain wave: It looks as if these lines, the trajectories of free-falling creatures, which really belong more to space-time than to the falling things, appear to be depicting the topography of a curved surface. That is, they are trying to tell us that this four-dimensional surface—space-time—is curved. In other words, it looks as if free-falling objects were visualizing for us the meridians, the skeleton, of a bent space-time, in the same way that you would be able to visualize the surface of a mountain by painting on it all the shortest paths taken by all possible hikers travelling on it.

Indeed, after many hits and misses, Einstein finally realized that one way to understand the effect of gravity upon free-falling bodies was to state that these objects follow lines called geodesics, "the

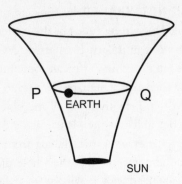

FIGURE 3.1   In the tubular space surrounding the sun, the Earth follows the shortest path between points P and Q. Whereas in flat space such a path is a straight line, in this space it is approximately circular. (This is a bit of a simplification, because the path of the Earth is a geodesic in *space-time*, and not in space only. Given that the Earth flows along the time dimension at the speed of light, its space-time trajectory is actually a spiral with a very long turn.)

straightest possible lines," on a curved space-time. And that gravity is nothing but this curvature of space-time. What a massive body such as the Sun actually does to its surroundings is to curve space-time. Free-falling objects then follow geodesics in this warped topography.

In Figure 3.1, I depict the reason the Earth follows a circle around the Sun. According to Einstein's conception, space around the Sun becomes tubular, as shown in the figure. To go around the tube in the shortest possible path you therefore have to travel in a circle—try other possibilities for yourself. Even though this is a bit of a cartoon for what actually happens, it gives you a taste for what is going on.

This turned out to be the right path through the labyrinth. It is an eccentric way to look at things, but it has many virtues. First, it allows you to get hold of a tool, called differential geometry, dealing with curved surfaces, precisely the horrible piece of mathematics I tried—and failed—to learn as a kid. Differential geometry is exquisitely well adapted to provide the language for this perception of the world, and when you use it to write equations for how matter produces curvature at a distance, you find that it is extremely easy to build into this "geometrical" action at a distance a speed of propagation—in fact, the speed of light. Thus we have a way of avoiding the inconsistency between gravity and special relativity. No longer is gravity an instantaneous action at a distance. Gravity is now the way in which mass curves space-time, an action that travels at the speed of light.

Another great virtue of conceiving gravity in this way is that it explains the mysterious equality of inertial and gravitational masses by doing away with these concepts altogether. Indeed, according to general relativity gravity is no longer a force, so bodies don't really have a weight or gravitational mass. But we do feel weight; if this feeling is not a force, what is it?

According to relativity, gravity is simply a distortion of space-time. In flat space, the law of inertia tells you that if no forces act on a body it follows a straight line, moving with constant speed; in other words, it suffers no acceleration. Einstein's theory states that under gravity, bodies are not subject to any force and so they again follow a straight line at a constant speed: in a curved space-time.

In this perspective, curvature takes care of it all; no longer is there a force of gravity. Therefore, no longer do the concepts of gravitational and inertial mass make sense, so that their identity is no longer a mystery. And yet if their identity in the Newtonian picture of the world were violated, no matter by how little, we would be unable to reinterpret gravity in Einstein's way and conceive it as geometry instead of a force. As it is, it all falls into place.

To sum it up, in this bizarre interpretation of gravity, matter affects the shape of the space around it, curving it. In turn, this curved space determines the trajectories of the objects moving through it: Matter tells space how to curve; space tells matter how to move.

All that remained to be determined was the exact equation detailing how matter produced curvature, which is nowadays called "Einstein's field equation." It was hard work, but all the conceptual difficulties had been overcome.

PEOPLE OFTEN WONDER how Einstein knew, after so much trial-and-error, when he'd got it right. It is sometimes said that "his sense of beauty" told him when he held onto the truth. This is only partly true. Sure enough, in 1915 he finally stumbled upon something that was too beautiful not to be true. But he had stumbled upon it before and abandoned it. In fact there were many objective and simple reasons that eliminated all the other possibilities, and to my mind these considerations are the most valuable of all. Certainly they are the most relevant to my own work on VSL.

In the early days, Einstein was led by the very obvious fact that the Newtonian theory is a very good description of all observations; as I mentioned before, space agencies still use it as a matter of course. Indeed, in 1915, apart from one very subtle exception (which I will relate presently), Newton's theory explained all known observations governed by gravity. Hence Einstein knew that whatever his general theory would turn out to be, whenever you did an actual calculation with it, you had to get a result very, *very* close to the Newtonian result. Luckily, this simple criterion ruled out numerous candidates. It was no longer a matter of finding the proverbial needle in the haystack. It was more like finding a haystack in a meadow.

This cunning approach reveals Einstein's eagerness to ride on Newton's back. It's easy to think that scientists just want to destroy everything everyone else has accomplished before, in the manner of certain other intellectuals. But this is usually not true in physics. Like rabbis, physicists always have to start by reasserting and praising all that previous physicists have asserted before moving on to point out subtle novelties, and this is how it was with Einstein and Newton's theories.

Nevertheless, the distinction between Newton and Einstein would be essentially a matter of taste if it were not for the fact that at a very subtle level they do predict different results. It is on this account that quite a lot of drama was in store—a drama in two parts, as we shall see. It touched upon an issue that mystifies nonscientists: Should science predict or postdict experiment?

I once gave a TV interview on VSL at the end of which I stated that at this stage I was just looking for experiments to decide whether the theory was right or wrong. The next day, a journalist accused me of not having helped matters by admitting that VSL was "just a theory"! In fact, most science is "just a theory," and is *not* motivated by existing observations crying out for an explanation. These "mere theories," however, should clearly *predict* new observations, new facts deduced by the theorist purely from mathematical calculations. If these predictions are observed, the theory is correct; if they are not, the theory is wrong. It's as simple as that. Science is not a religion.

The idea behind predictions is that the responsibility for telling observers what to look for should fall upon theorists. Trying to expand our knowledge by waiting for new observations to be found by accident is like shooting in the dark. There are so many possible directions; how do we know where to look to find something new? Better to have a guiding theory telling you what to look for. Undoubtedly, it *is* observation that establishes facts, but without a theory one risks wasting a lot of time looking in vain.

Naturally, science sometimes proceeds in the opposite way, and if that happens, so much the better. Experiment may get ahead of theory, so that we come across new facts first by observation. Theory is then about *postdicting* existing observations. The role of the theorist is now to collect existing new data and come up with a theory "explaining it all"; that is, the theorist must find a framework within which all observations make sense.

Therefore, in reality, both prediction and postdiction play an important role in science—they are not mutually exclusive. And indeed Einstein's views on gravitation were vindicated by two stunning observations: the first a postdiction, the second a prediction.

IN 1915, THERE WAS a single phenomenon that Newton's theory of gravity could not explain. Planets describe nearly circular orbits around the Sun, but closer inspection reveals that their orbits are in

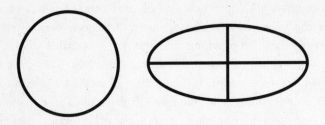

FIGURE 3.2   A circle (left) and an ellipse (right) with its axes depicted. The more one axis differs from the other, the more an ellipse departs from a circle—and the higher its "eccentricity."

fact very "circular looking" ellipses. Figure 3.2 depicts this geometrical object, exaggerating heavily how it differs from a circle. The ellipse has two axes, also depicted, and the larger the difference between their lengths the more the ellipse departs from a circle, or, in mathematicians' jargon, the larger its *eccentricity*.

With the exceptions of Mercury and Pluto, the orbits of the planets in our solar system are not terribly eccentric. For instance, the Earth's orbital axes differ by only a couple of percent, so that our distance to the Sun never varies by much. Nonetheless, the departure of planetary orbits from circles is well within the reach of astronomical observations and was quickly established in the wake of the Copernican revolution (a posh expression for putting the Sun, rather than the Earth, in the center of the solar system). The elliptical nature of these orbits was first inferred from astronomical observations by the mathematician Johannes Kepler, in what is now known as Kepler's first law.

Kepler's law, to some extent, can be seen as a postdiction of Newtonian theory. Indeed, in his famous book *Principia*, Newton derives Kepler's law mathematically. However, Newton's derivation assumes that the solar system contains only the Sun and one planet (which can be any planet). But in reality there are several planets in

FIGURE 3.3

the solar system, so that each planet is subject not only to the Sun's gravitational pull but also, to a much lesser extent, to the pull of all the other planets. One must therefore refine this approximation in Newton's original calculation, and the way to do it is to say first of all that planets know about the Sun and nothing else and therefore follow elliptical orbits; then that they are perturbed by all the other planets so that their elliptical orbits are changed accordingly. All we need to calculate is this small correction.

It's a typical physicist's calculation and the result, using Newtonian theory, is that due to the perturbing effects of all the other planets, each ellipse should rotate around itself very slowly; that is, its larger axis should slowly change direction as the planet itself travels around the ellipse much faster. Hence the exact Newtonian prediction for the orbits of planets is a rosetta such as the one depicted in Figure 3.3. The effect is very small, with the failure of each near-elliptical revolution to close being quite imperceptible, so that each "year" the planet covers territory only slightly shifted from the previous one. The full rotation of the ellipse usually takes thousands of the planet's years.

This phenomenon was indeed observed in the nineteenth century, at a time when all the planets up to Uranus had been discovered, and excellent agreement was found with Newtonian calculations. This included the orbits of Venus, Earth, Mars, Jupiter, and Saturn. However, some discrepancies between theory and observation were found for Uranus. When one computed the perturbing effects from all the inner planets (Neptune and Pluto being still undiscovered) one did not quite find the observed rosetta. Something seemed to be missing, either from the theory or the observations.

At this point we may taste a delicate masterpiece of prediction, cooked up by the French astronomer Urbain-Jean-Joseph Le Verrier. Such was Le Verrier's faith in Newton's theory that he allowed himself to do something quite radical. He decided that a way out of the impasse was simply to postulate the existence of an outer planet with a perturbing effect upon Uranus capable of explaining, according to Newton's theory, precisely what had been observed.

This planet, dubbed Neptune, would be so far away from the Sun that it would be very faint, explaining why astronomers had not seen it thus far. However, Le Verrier went on to compute several properties required of this hypothetical planet; in particular, he told astronomers where and when they should look for it. A few years later, Neptune was duly discovered, precisely where and when Le Verrier said it would be. Impressive.*

This incident did a lot to establish even more firmly Newton's theory of gravity. But soon enough, another planetary anomaly was discovered, this time in Mercury's orbit. Mercury's ellipse is unusually eccentric and rotates around itself faster than that of other planets. Still, the period required for Mercury's ellipse to complete a full revolution is about 23,143 terrestrial years. This is not to be confused with Mercury's year, which is the time it takes to travel around the ellipse itself, and is only 88 terrestrial days.

---

*Pluto is far too small to have any effect upon Uranus or Neptune.

However, calculations using Newton's theory, even taking into account the perturbing effects of all other planets, led to a different figure: Mercury's ellipse should complete a full revolution around itself in about 23,321 terrestrial years. Somehow Mercury's ellipse is rotating slightly faster than the Newtonian prediction. Yet again scientists concluded that something was missing, either from the theory or the observations.

Not surprisingly, in view of his previous success, Le Verrier decided to play his trick again, and this time he postulated the existence of an inner planet, Vulcanus. Vulcanus would be smaller than Mercury and would be very close to the Sun, so that observing it would be extremely difficult. For one thing, it would be very faint, for another it would always be seen very close to the Sun, possibly never at night. This would explain why it had not been seen before. Again Le Verrier computed where and when astronomers should look for Vulcanus, and prepared himself to receive a second round of applause.

But when the search for this new planet took off, the results were very disappointing indeed. Vulcanus was never observed. The years went on, and from time to time amateur astronomers in search of their moment of glory would "observe" the elusive planet, but none of these claims were ever independently verified. Vulcanus fell into the territory nowadays occupied by UFOs—if you desperately wanted to see it, it would appear to you, but no watertight scientific detection was ever made. Scientists did not quite know what to make of this. It became a mystery people decided to live with rather than to explain.

Imagine Einstein's joy when he realized that applying his final theory of gravity to the orbit of Mercury accounted exactly for the observed rosetta without the need for Vulcanus! The correction imposed by his theory upon the Newtonian calculations was sizable for Mercury, but negligible for all the other planets. And so his theory took on board all the successes of Newton's theory while solving the one problem Newton's theory couldn't. It could not have been any better.

According to his own accounts, for a few days Einstein was beside himself with excitement, unable to do anything, immersed in an enchanted, dream-like stupor. Nature had spoken to him. I have often said that physics is fun because it can give you a massive adrenaline rush. What Einstein got at this moment must have been the ultimate overdose.

BUT A SECOND APPROVING NOD from nature waited for him, this time involving the perilous territory of predictions. Right from the start of his reflections Einstein had concluded that if Galileo's leaning tower experiment was to be taken seriously, light should also fall under the effect of gravity. If gravity does not look at who is falling, light should behave under gravity in much the same way as other fast-moving objects. The latter see their trajectories bent by gravity; the more so the slower they move. Therefore, light rays should bend near very massive bodies, if only very slightly. The question now was by how much.

It turned out that the answer varied widely from theory to theory, even when one considered only theories that reduced to Newtonian predictions in a rough approximation. The first calculations of this effect were performed by Einstein around 1911, in fact within the context of his varying speed of light theory. To maximize the effect, so that there might be a chance astronomers could observe it, he considered the following set-up.

He first looked for the most massive object around us, given that the larger the mass the stronger the gravity, and therefore the more light would bend. He therefore elected the Sun as the source of gravity.

He then considered light rays that would just graze the Sun. Since he knew that the effect of gravity decreases very quickly with distance, the closer the light ray got to the Sun, the more it would bend.

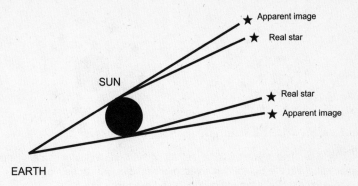

FIGURE 3.4   Einstein's set-up for detecting the deflection of light rays by the Sun: an observer on Earth, looking at a star whose light just grazes the Sun. Due to light ray deflection the apparent image is shifted outwards.

He considered therefore what would happen to the image of stars appearing in the sky just off the Sun's disk, or more specifically he computed how their apparent position in the sky would change because of light-ray bending.

But surely no one can see stars near the Sun, because if you can see the Sun it's not nighttime! Well, not quite. That is what eclipses are for, at least as far as astronomers are concerned. During a total eclipse, the disk of the Moon completely covers the disk of the Sun, so that one can see the stars just around the Sun in an exotic night sky, realized during the day.

Einstein's set-up is therefore like the one sketched in Figure 3.4. As the picture shows, the Sun's gravity should act like a subtle and gigantic magnifying glass spreading images outwards, and indeed the term *gravitational lens* is sometimes used to describe this phenomenon. It is even possible to see stars "behind" the Sun, because light rays do go around corners, as long as the corners are sufficiently massive.

Of course the effect is very small, so this is a subtle experiment requiring the expert use of a few extra tricks. The most obvious one is to look for clusters of stars rather than a single star, and to inspect how their relative positions might appear to be distorted due to light

FIGURE 3.5   The gravitational-lens effect of the Sun. A cluster of stars (left) passing behind the Sun during an eclipse would appear to be magnified (right). (This picture is grossly distorted and exaggerated.)

rays' bending as they pass behind the Sun. Many such groups of stars can be found in the sky, and with a bit of luck one could be found just behind the Sun (as seen from Earth) during a total eclipse. All one had to do was to take two sets of photographs of the candidate cluster: one well away from the Sun, and one during an eclipse, the starlight just grazing the Sun. And then compare the plates. The latter should appear expanded as if magnified through a lens (see Figure 3.5).

This was the task supplied by Einstein to astronomers; but he still had to compute the exact deflection angle according to his general theory of relativity. Einstein's 1911 calculations had already produced a concrete prediction in this respect. He found that light rays just grazing the Sun should be deflected by approximately 0.00024 degrees. That's about the angle subtended by a football 50 km away from you—very small indeed. For reference, the angle subtended in the sky by the Sun, or the Moon, is about half a degree. However, the top telescopes available at the beginning of the twentieth century could just resolve that sort of angle. Therefore the effect predicted by general relativity could in principle be observed.

The problem was finding the right conjunction of conditions. Although total eclipses are rare if you confine yourself to one region of the Earth, overall they are actually quite common. At least two solar eclipses must happen every year, the maximum number being five. However, not all of these are total eclipses; that is, it may be that the Moon only partly occludes the Sun. Furthermore, not all total eclipses are eligible because it is a fluke to find the Sun and the Earth aligned with a cluster of stars while, perchance, they are also aligned with the Moon. The two events are totally unrelated: It's like having a full Moon on Friday the thirteenth. Therefore astronomers had to wait patiently until the right conditions came along before they could finally put the theory to the test.

Which was just as well. As it happens, the deflection angle computed by Einstein in 1911 was wrong. As I pointed out earlier, the calculation of this effect is highly dependent upon the fine details of the theory of gravitation you are using, and all but the final formulation of general relativity led to the wrong result. The actual prediction of general relativity in its final form is exactly twice the value mentioned above: 0.00048 degrees of deflection for a light ray just grazing the Sun, instead of 0.00024 degrees. Therefore, until 1915, when Einstein found his final theory, he had made the wrong prediction for this effect. How very embarrassing.

Just to make matters worse, as fate would have it, there were two eclipses with perfect conditions for observing the gravitational lens effect between 1911 and 1915, and astronomical expeditions were promptly sent to the regions where they could be observed.

The first expedition was led by Argentineans and made use of a promising total eclipse expected in Brazil in 1912. A nice, highly populated swarm of stars would be seen behind the Sun at the time, creating perfect conditions for the experiment. Plates of this cluster of stars while away from the Sun were prepared and the expedition set off full of hope. A pity it rained heavily all day and there was nothing but cloud cover to see.

The second expedition was led by Germans, in 1914, and preyed upon an eclipse that could be seen from Crimea. The astronomical tables once more revealed perfect conditions for the observation of the effect, indicating that a rich cluster of stars would be seen around the Sun at the time of the eclipse. Again plates of this cluster away from the Sun were prepared, and the expedition departed in high spirits. All seemed to go well, and indeed the weather was rather fine, when a few days before the eclipse World War I broke out, and the expedition suddenly found itself in enemy territory. Some fled in time; others were arrested. They all eventually made it home safely, but needless to say, without any data.

A lucky star, therefore, seemed to look over Einstein's shoulder as he made mistakes, corrected them, and slowly made progress towards the final version of his theory. Astronomers accidentally granted him a reprieve while he worked out the finer details.

It was not until 1919 that a British expedition led by Eddington and Crommelin finally made an observation of the effect. By then, Einstein had found the final form of general relativity, with the correct prediction, which was duly verified by the observations.

The lucky bastard!

4   HIS BIGGEST ERROR

I LIKE TO THINK OF THE UNIVERSE as an organic being, something alive. We are all cells of this being, all the suns we can see in the sky shed light making up the blood flowing through the universe's immense cycles. The forces governing this unique being are physical forces, just like the forces that control and make human beings. And as is the case for each one of us, when we look at the big picture, we see that the individual vastly transcends the mechanism controlling the bits and pieces that make up the whole.

Einstein's next enterprise was nothing less than a mathematical model, based on general relativity, for this gigantic beast. The model described the universe as an unusual substance called the *cosmological fluid*. This fluid was made up of extraordinary molecules—nothing less than whole galaxies. But he soon found that his gravitational field equation allowed him to work out the relationships between all the variables describing the universe, as well as how these variables changed over time. When he did this, however, Einstein got an unpleasant shock. His equation suggested that the universe was restless rather than static. According to general relativity, we should live in an expanding universe that exploded from a violent birth into a Big Bang cradle.

In some ways, the restless universe revealed by general relativity is indeed like some people—a wild, bad-mannered, untamed beast. Except that the temperamental universe derives its restlessness from a simple hormonal problem: Gravity is attractive. This is true whether we picture gravity as a force (following Newton) or as geometry (following Einstein). It's common sense. The earth pulls us toward its center rather than kicking us into the skies.

However, this simple fact, the attractiveness of gravity, is sufficient to suggest the impossibility of a static universe—a fact Einstein noted at once. Here's how the argument goes. In your mind, conjure up such a static universe and let it evolve. Left to its own gravity it would promptly collapse under its own weight, each of its parts attracting all others in a contracting motion that would terminate in a Big Crunch. The only way to prevent gravity from leading to a collapsing universe is to allow for an expanding universe, in which everything flies apart. Then gravity slows down the cosmic expansion, its attraction pulling everything back together, decelerating the cosmic rush. But if the outward motion is fast enough, the gravitational pull will never halt it, and thus a Big Crunch can be avoided.

To be more precise, if we allow for a universe in outward motion we have two factors at war: the cosmic motion and the force of gravity. We must then weigh how fast the universe is expanding at a given time against how much mass it has (and so how strongly gravity is pulling it all back together). Out of the scales comes a crucial speed for a given mass—an escape velocity for the universe. This is not dissimilar to what happens when a rocket tries to leave the Earth. Give it a sufficiently high launch speed and it eventually escapes Earth's gravity and flies away forever. Too sluggish a start and the attractive force of gravity will eventually bring it back to Earth. Likewise, for a given matter density in the universe there is a critical cosmic expansion speed below which the universe eventually stops expanding and falls back on itself, but above which it expands forever.

In all possible scenarios, simply because gravity is attractive, the universe does not want to sit still. It stubbornly wants to move, either to expand or to contract, something Einstein refused to believe. And here starts his terrible error: the struggle to find a static universe within his field equations.

IN 1917, THE PERMANENCE of the universe was a fixed belief in Western philosophy. "The heavens endure from everlasting to everlasting." Thus, it disturbed Einstein greatly to discover that his field equation predicted a nonpermanent universe. Faced with this contradiction between his theory and the firm philosophical beliefs of the day, Einstein weakened. He modified his theory.

Perhaps if he had been just a little bit dimmer he never would have blundered. He would not have been able to find a way to fix a problem that did not exist, and would eventually have accepted what his own mathematics was trying to tell him. As it was, he was too clever for his own good, and soon found a simple modification of his field equations that allowed him to build, in his mind, a static universe.

He did this by introducing another term into his field equation, the so-called Lambda (after the Greek letter he used to denote it), often referred to as the "cosmological constant." It was an abstruse modification, essentially amounting to ascribing energy, mass, and weight to the *nothing* or *vacuum*. It was also an ugly fudge factor in an otherwise beautiful theory, something that was introduced arbitrarily for no other reason than to ensure that a static universe could be predicted by the general theory of relativity.

The cosmological constant is a simple modification of Einstein's field equation, and at first looks quite innocuous. Predictions for the orbit of Mercury and for light deflection, for example, remained essentially unaffected. But for the purpose of cosmology, and at a very fundamental level, the story was now totally different. The cosmological constant is the 666 of physics, to this day an ugly beast we do not seem able to shake off. While working on VSL, I lost many a night's sleep, haunted by the Lambda bogeyman.

Like all devilish things, the early days of the cosmological constant were innocent enough. As we know, according to general relativity everything falls democratically, in the same way, along lines in space-time called geodesics. The reverse of the coin is that everything also produces gravity, that is, everything curves space-time and bends geodesics. This fact implies some surprising new effects, far removed from our experience, but known to be solid predictions of relativity from the beginning. For instance, light and electricity are heavy. Not only is light bent by gravity, it also attracts other objects; a sufficiently energetic light ray would pull you towards itself. Motion is heavy, too, and a fast star attracts others more than does a steady one. Indeed, gravity emanates from everything, be it heat, light, magnetic fields, *even gravity itself*. It is this last feature that makes the mathematics of relativity so complicated: It describes matter producing gravity and then gravity itself as the source for more gravity in an intricate cascade.

That much was already obvious from Einstein's standard field equation. But he then asked a penetrating question: Could "nothing"—the void—also produce gravity? And if so, what is the weight of the nothing?

This may sound like a nonsensical question, but we have seen him asking crazy things before, with devastating consequences. And this question does not really come out of the blue. In fact Einstein's dealings with the "nothing" were always complex, and to some extent this question, and the genesis of the cosmological constant, was the climax of a long and contorted relationship.

There was once a time when scientists believed that "something" pervaded the "nothing." That something was dubbed the "ether," a scientific counterpart to the ectoplasm. The ether theory reached the height of its popularity in the nineteenth century, along with the so-called electromagnetic theory of light; and although the ether may sound like a strange concept today, a moment of thought reveals that it is, a priori, quite sensible.

The argument for the ether went as follows: Light is a vibration, or a wave; that much was well understood at the time, and was supported by a large body of evidence. All other vibrations—sound waves or ripples in a pond, for example—need a medium to support them, something that actually vibrates. If we remove air from a container with a pump, no sound can propagate through the container because there is nothing to vibrate in the form of sound. Ripples on a dry pond don't make any sense whatsoever.

But if we pump everything out of a box and create a perfect void, light still propagates through it. Indeed, there is just such an excellent void through interplanetary space; nevertheless, we can see stars twinkling in the sky. It's as if when you pumped everything out of your box you forgot to remove something that could support light's vibrations, or as if the interplanetary void were actually filled with a similar substance. That something was the ether, a subtle substance that pervaded everything, the existence of which we could infer only from light itself. We could not touch it or feel it, or even pump it out of a container; and yet, as light propagation attested, this ethereal substance was omnipresent. The ether was therefore believed to be as much a part of reality as any other element, and one may find it inscribed on the margin of most nineteenth-century periodic tables.

The ether was killed by Einstein's special theory of relativity because it contradicted the constancy of the speed of light: An ether wind would accelerate or decelerate the vibrations it supported, that is, light. It was this ether wind, rather than mooing noises in a dream meadow, that motivated the Michelson-Morley experiments. As the Earth moves through the ether, we should be blown by an ether wind coming from different directions (depending on our direction of motion), something that should translate into a change in the speed of light (depending on light's direction with respect to the wind).

If you accept the reality of the ether, then the negative result of the Michelson-Morley experiment, the constancy of the speed of light, is totally nonsensical. How can observers at motion with respect to each other have the same relative speed with respect to the ether? If the constancy of c by itself is baffling, within the ether theory it makes no sense at all.

This conundrum led to all sorts of desperate proposals. Some noted that the Michelson-Morley experiments were always carried out in basements, where laboratories are usually situated. It was suggested that perhaps the ether got stuck inside basements, and so we could not feel its wind. It's a perfectly ridiculous solution: If we could not feel the ether in any other way how come basements had the ability to trap it? Surely if the ether got trapped inside basements one ought to be able to trap it inside containers—or pump it out of them. Nevertheless, people did try to repeat the Michelson-Morley experiments, in the hope of detecting a change in the speed of light on the top of mountains, where the possibility of ether-trapping was naturally excluded. To no avail; the ether winds were never detected.

Einstein was the first to propose that light was a vibration without a medium, a ripple *in vacuum*. Without this conceptual leap, the special theory of relativity would never have been possible. Indeed, if you didn't find special relativity too hard a concept to swallow, it is perhaps because you were never taught the concept of ether in school.* Since Einstein's 1905 breakthrough, the ether has become the preserve of historians of science—and the few scientists who know about it ridicule it. Yet the ether was the main mental block that delayed special relativity, and getting rid of it was a large part of Einstein's genius. In his own words, in his seminal 1905 paper: "The introduction of a 'luminipherous ether' will become superfluous in our theory, since we shall not need the concept of 'space at absolute rest.'"

He thus restored nothingness to the nothing, voidness to the void. And now, some twelve years later, in the middle of a cosmic affliction, he was reversing himself completely, asking whether one could ascribe some sort of existence to the void after all, so that the void could produce gravity. Could nothing be something?

---

*I first came across the concept of ether during my precocious reading of *The Evolution of Physics*. When I questioned my physics teacher about it, he told me not to be stupid, stating that "if everything were pervaded by ether we would all be anaesthetized."

WHILE EINSTEIN LIVED in Bern, working as a patent office clerk, he did his research work in a small study away from his home. In this study he kept a large number of cats, of which he was very fond. However the cats at times could be rather burdensome, scratching persistently at closed doors, demanding to roam freely throughout the house. He could not leave all the doors open, so he decided to cut holes in the bottom of the doors, producing cute little cat doors.

In that year he had roughly equal numbers of large and small cats. Therefore, quite logically, he cut out two holes in each door: a large one for the large cats, and a small one for the small cats. It made perfect sense.

One may gather from this that Einstein's contorted mind already demanded that "nothing" be "something." A hole should have a meaningful existence, and the small cats might be offended if a personalized nothing was not prepared for them. If you are ready to go down this surreal path, then perhaps the rest of the argument will feel natural to you. Indeed, in a similar vein, Einstein ascribed existence to the nothing, proposing that the vacuum could produce gravity. But as he worked out a consistent way for this to be possible in his theory, he found a curious result: The vacuum should be gravitationally repulsive. At this point he must have jumped up in the air, for he knew that the impossibility of a static universe resulted directly from the *attractive* nature of gravity. Was *repulsive* vacuum energy the way out?

The proof of the repulsiveness of the vacuum stems from well-established mathematical results within general relativity. According to relativity, the attractive power of a body is a combination of its mass and its pressure. Compress an object and its ability to attract other objects is enhanced. The Sun is under pressure, therefore its ability to attract planets is greater than it would be if it were a pressure-free dust ball. The effect is actually very small because in ordinary objects, and even in the Sun, the amount of pressure is far outweighed by the amount of mass. But the effect is clearly predicted by general relativity, and if one could supercompress an object, one should be able to observe it.

This much was not controversial, and was an intrinsic part of the predictions of relativity. But now one notices an interesting thing: Tension is a negative pressure, and so the effects of tension should be such that the attractive ability of objects is reduced by its presence. A stretched rubber band attracts less than one would expect from its mass or energy content alone. A hypothetical tense Sun would also lose some of its attractive power.

Again the effect is very small in normal objects, but in principle there is nothing stopping us from increasing the amount of tension in a body so much that gravity becomes repulsive. Therefore, according to relativity, gravity does not need to be attractive. To create repulsive gravity all you need is to find something truly on the edge, overcharged with tension, ready to snap.

Like what? The surprise is that the vacuum just might be a very good example. When Einstein worked out a mathematically consistent way to give the vacuum a mass (i.e., energy—$E=mc^2$) he found that he could not avoid giving it a very high tension. It's a bizarre fact, but it pops out of the only possible equation consistent with differential geometry that allows for a vacuum energy.

The vacuum tension is very high, so much so that the gravitational effects of tension overcome those of its mass and, as a result, the vacuum is gravitationally repulsive. In Newtonian language, the void has negative weight.

Naturally, the void's energy is very diluted and uniformly spread throughout everything. On the scale of the solar system, the gravitational effects of matter far exceed those of the vacuum. One has to go to cosmic distances for the vacuum density to become comparable to that of ordinary matter, and for the repulsive side of gravity to become apparent.

TO SUM THINGS UP, Einstein knew that a restless universe was an immediate implication of the attractive nature of gravity. But now he also knew that, with a cosmological constant, gravity need not be attractive. The question was this: Could one cook up a static universe by making judicious use of the new ingredient?

Einstein's recipe was as follows. Take a model expanding universe that does not quite have the escape speed. Eventually, its gravity overcomes its expansion, and it collapses back in on itself, culminating in a Big Crunch. In your mental construction take such a universe as it is about to stop expanding and start collapsing, so that the universe is actually momentarily steady, and sprinkle it with a carefully measured amount of cosmological constant. Because this vacuum energy is gravitationally repulsive, it pushes out against the attracting effect of normal gravity. One type of gravity wants the universe to contract, the other for it to expand. If you used the ingredients in the right proportions, the amount of attraction may just cancel the amount of repulsion, and the universe sits still.

And so Einstein managed to contrive a model for a static universe within the theory of relativity, albeit only with the aid of the cosmological constant. Truth be said, the universe is not happy about its stillness; indeed, it looks as if it were in a straitjacket, an imposed but unstable quietude. But it does sit still, for the benefit of generations yet to come.

Thus the universe complies with what was just semireligious prejudice, a belief upheld and taken for granted within the context of Western culture. Ironically, just as cosmology was about to escape the grip of religion and philosophy, the latter took revenge and poisoned the first scientific model of the cosmos. To give Einstein credit, science is based on data, and at the time there was no cosmological data; so prejudice took its place. The recipe he found to accommodate this prejudice is ingenious, and we might never have known about the cosmological constant without it. Thus he found what is now called the Einstein static universe—his biggest blunder.

BUT SOON ENOUGH, astronomical data about the universe started to pour in. During the 1920s, the astronomer Edwin Hubble carried out a series of groundbreaking observations from Mount Wilson, California, which quickly became the best lookout into the universe of its time. In its heyday, Hubble's telescope was so famous that Hollywood stars begged to be allowed to look through it. The universe had become fashionable.

Hubble had been trained as a lawyer, but he soon saw the error of his ways and decided to devote himself fully to astronomy. Yet his career was not quite that of a scholar, either. He excelled mainly as a sportsman, in basketball, boxing, swordsmanship, and shooting.* The latter was of great assistance to him when he had to fight a duel with a German officer who, after his wife fell into a canal, challenged Hubble for rescuing her. Always an anglophile,† he studied at Oxford, where he seems to have absorbed from the English culture a leaning towards eccentricity, as demonstrated by his idiosyncratic astronomical observations. In these he was helped by another self-taught observer, Milton Humason, a man whose fascination for astronomy led him to join the Mount Wilson staff as a teenager (originally in charge of the mules that brought equipment up the mountains). The two men shared backgrounds unbefitting to professional astronomers, but they had great enthusiasm for their work, and unbeatable flair. Together, they would change the perspective of cosmology.

Perhaps because Hubble had little training in the field he made very unusual observations. He installed a telescope inside a building that rotated as a perfect clock, moving exactly to counter the Earth's rotation. He could thus automatically point his telescope in the same direction for extended periods and make observations without "attaching" the naked eye to the end of his telescope, using instead photographic plates that could be exposed for very long periods.

What came out of these unusual observations was truly pornographic. In Figure 4.1, I show you a picture of a galaxy—that's an island of stars just like our Milky Way, which probably does not

---

* According to John Gribbin, Hubble was not a great athlete but a great liar.
† He was an insufferable snob, disliked by all.

FIGURE 4.1   A galaxy.

shock you. But before Hubble, no one had ever seen a galaxy—a swirling spiral of billions and billions of stars encircling a bright central eye—and naturally enough people *were* shocked. It was as if someone had invented a new camera, and the first time a photo was taken with it, it was found that we are actually surrounded by otherwise invisible little green men, happily living amongst us.

In fact, galaxies are not particularly small in the sky; the largest ones actually have an apparent size comparable to that of the Moon. They are too faint for our eyes, however, even through a telescope. Only Hubble's clever trick could dig them out of the darkness of the skies.*

The discovery of galaxies was to radically change the perspective of cosmology, showing how badly misguided most theorists' efforts had been thus far. If we look at a clear night sky with a trained naked

---

* Hubble was not the first to use astrophotography or to sight galaxies, but he was the first to understand their cosmological significance.

eye, we can see an overwhelming plethora of detail: planets, stars, our own galaxy—the Milky Way—and, if we catch a glimpse of the Magellanic Clouds, also a faint image of a satellite to our galaxy. We see so much detail that, from this perspective, the task of predicting the behavior of the universe as a whole seems nearly impossible, akin to predicting the Earth's weather or the path of ocean currents from one small point on the planet.

Hubble's discoveries, however, show us that all this is irrelevant detail. With the aid of very good telescopes we can zoom out to find that the stars in the sky are indeed part of the galaxy called the Milky Way. We find that this galaxy is one of many similar "islands of stars" floating about in the universe. Zooming out further, we see that most of these galaxies tend to pack in groups or clusters.

If we zoom out even further, however, the picture changes dramatically: We start to see that all these structures, galaxies, clusters of galaxies, even the largest structures we can see, are constituent particles of a rather boring soup, the cosmological fluid. In contrast with the complex diversity of our local neighborhood, this soup looks extremely uniform. It depicts for us a very homogeneous universe, completely devoid of structure. And surely such a tame object is simple enough for physics modeling. But the crucial point is to recognize that the basic units, the "molecules" as I called them earlier, of this very simple fluid are huge and invisible to the naked eye: They are galaxies, not stars or planets or any of the "trivia" that can be appreciated without the aid of a telescope.

This was the first blow Hubble dealt upon cosmologists. Specifically, Hubble taught cosmologists that the study of the universe only made sense if one first took into account its enormous size, much the way one cannot understand or appreciate the plot of a film if one's eyes are placed only a couple of inches from the movie theater screen.

This revelation opened the doors to cosmology—and the task of explaining the universe became much simpler. But Hubble also discovered something else, something far more intriguing and even more far-reaching. He found that this uniform magma seemed to be expanding, revealed by the fact that all the galaxies we can see appear to be moving away from us. Hence, seen from a suitable vantage point, the universe is not static after all! Einstein must have blushed upon hearing this news. Had he stuck by his original equation, and accepted the conclusions that it seemed to require, he could have *predicted* that the universe must be expanding, and carried off the greatest scientific coup of all time.

The way galaxies move away from us displays a distinctive pattern. It satisfies Hubble's law, which states that the speed of recession of a galaxy is proportional to its distance away from us. A galaxy that is twice as far away from us as a second galaxy moves away from us at twice the speed of that second galaxy.

At once we may deduce that Hubble's law has a disconcerting implication. If we see the stuff of the universe receding from us faster the further away it is, then a massive cataclysm must have happened in the past. To see why, let us rewind in our minds the film of the universe and look at it backwards in time.

If a given galaxy is moving away from us in the real film, then in the backwards film we see it moving towards us. That means that at some time in the past, that galaxy must have been on top of us. How far into the past do we have to go to witness this horrifying state of affairs? Well, that's the current distance of the galaxy, call it L, divided by its speed, call it v: The rewind time for collision is L/v.

If you think that this single cosmic collision looks calamitous, now ask yourself the same question for any other galaxy, say a galaxy at twice the distance. According to Hubble's law, its speed is now 2v, so the rewind time for collision is $2L/2v$, which is the same time as before: L/v. Therefore the second galaxy will be on top of us at exactly the same time as the first one (see Figure 4.2). And so on and so on. Galaxies farther away from us have a longer distance to travel before they collide with us in the backwards film. But they also move

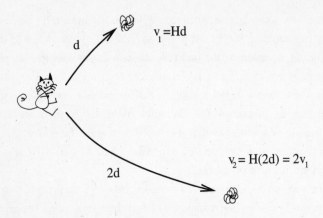

FIGURE 4.2   An observer contemplating the Hubble flow of two galaxies at distances d and 2d. The second galaxy recedes at twice the speed of the first. Hence the observer infers that they were both coincident with himself at the same time in the past. The conclusion is valid for all galaxies, and so the whole universe must have once been compressed in a single point that exploded in a Big Bang.

faster. The whole thing is structured so that the collision time is the same for all galaxies in the universe.

Looking again at the film in forward motion, with the arrow of time pointing in the right direction, we therefore arrive at an amazing conclusion. Hubble's law implies that at some time in the past the whole universe was concentrated in a single point. It then appears that the entire universe was ejected from this point in a huge universe-creating explosion. More concretely, from the observed recession speeds we can estimate that the Big Bang happened some 15 billion years ago. The expanding debris of this explosion compose the universe we can see today.

Surprising as this conclusion may be, the thinking behind it is hardly unique to the dynamics of the universe. If you look at the debris of an exploding grenade, you find that it, too, satisfies Hubble's law. Hubble's law is the hallmark of any big explosion.

In effect, by discovering his famous law, Hubble had just found evidence for the Big Bang.

5 THE SPHINX UNIVERSE

AS REVEALING AND AS REVOLUTIONARY as Hubble's discoveries were, it eventually became apparent that not all was a bed of roses with the Big Bang theory of the universe. Although the model did not conflict with any known fact—and to this day has proven impregnable when confronting observations—some features of the universe remained unexplained. This led to puzzling questions, among them: Why does the universe look the same over such vast distances? Why is the universe so large? Indeed, why does the universe exist at all?

One may list some half-dozen similar difficulties, collectively known as the cosmological problems or, more colloquially, the Big Bang riddles. These puzzles have motivated much controversy and radical work in recent times. Trying to answer them does not mean abandoning the Big Bang universe as we see it nowadays; for we know that such a model can no longer be disputed. Instead, in a strange Freudian twist, cosmologists are searching the Big Bang's infancy for clues to its adult behavior. They hope to replace the actual bang, or perhaps what happened in the tiniest fraction of a second after it, with something less extreme, giving the universe a less traumatic birth and childhood. Such an operation does not conflict with observations because we have no direct access to such early stages of the universe. However the answers to all the riddles might perhaps lie in this reformed bang.

When I became a physicist, in Britain in 1990, the Big Bang riddles were still up in the air, and they presented a major challenge for any young apprentice cosmologist. Above all, they showed that the field was ripe for innovation. There were deep questions yet to be answered, awaiting our speculations and imagination to address them. I recall being surprised to find, after all the hype associated with the Big Bang successes, that there was still creative work to do in cosmology. While attending graduate-level classes at Cambridge, I started feeling that these riddles alone were sufficient reason to prefer cosmology over the other two exciting areas in frontier physics: string theory and particle physics. Of these, the first has no data, only speculations; the latter has too much data and seemed to me to have little room left for truly creative work. Cosmology was just right—well founded in reality, but still immature enough for fundamental problems to remain unsolved.

THE SIMPLEST BIG BANG riddle is called "the horizon problem" thus named because cosmological observers can see only a small portion of the universe. They are surrounded by a horizon, beyond which they cannot see. In our daily experience we know that we cannot see the whole Earth, but only what is within our horizon. The inhabitants of a Big Bang universe suffer from a similar perspective problem, except that what causes the horizon effect on Earth is its curvature, whereas the source of horizons in the universe is a combination of two very different phenomena. The first is the existence of a universal speed limit, the speed of light. The second is the fact that the Big Bang universe has a birth date and therefore a finite age at any given time. Combine these two facts and you'll see that they lead immediately to the prediction of horizons: Creation entails limitation.

When you look up at the sky, at a distant star, you see it as it was in the past. You see light that left the star a long time ago, then spent all the time since traveling towards you. Some stars in the sky are about 1,000 light years away; that means you are seeing them as they were 1,000 years ago. Over the past thousand years the image you see was busy traveling from that distant star all the way to you.

But now let's go megalomaniacal and look farther and farther away, into the vast distances probed by astronomers since Hubble. The farther away you look, the longer the time delay between the production of the image you see and your seeing it. Therefore, as you probe deeper into the distance, you are also probing deeper into the past. As we look at galaxies a billion light years away from us, we see them as they were a billion years ago. We see shadows of their past, and perhaps they don't exist anymore—we shall never know.

This game places cosmologists in a better position than archaeologists: We have direct access to the universe's past; all we need to do is look far enough away. But the game also leads to a rather disquieting conclusion. As we look farther and farther away, eventually we reach distances for which the look-back time is comparable to the age of the universe, 15 billion years. Beyond such distances, on the order of 15 billion light years, we clearly cannot see anything: These distances establish our cosmological horizon. This is not to say that regions beyond the horizon do not exist; they certainly do. However, we cannot see them because none of the light they have emitted since the Big Bang has yet had time to reach us.

If light traveled at an infinite speed, there would not be a horizon effect. Similarly, if anything could travel faster than light we would be able to learn from regions beyond the horizon if they emitted signals via the faster-than-light channel. Finally, if the speed of light was not a constant, and light could be accelerated, say by moving its source, then we should be able to see objects outside the horizon, so long as they were moving towards us sufficiently fast. It is the fact that the speed of light is a finite constant, which acts as a universal speed limit, that imposes the horizon effect upon any universe with a finite age.

The existence of horizons in the universe is not by itself a problem. The problem is the size of the horizon immediately after the Big Bang. When the universe is one year old, the radius of the horizon is only one light year. When the universe is one second old, the radius of the horizon is measured by how far light can travel in one second: about 300,000 km—the distance from Earth to the Moon. And the closer we get to the Big Bang, the smaller the horizon becomes.

The baby universe is therefore fragmented into tiny regions that cannot see each other. It is this shortsightedness of the early universe that really gets us into trouble, for it precludes a physical explanation, that is, one based on physical interactions, for why the early universe appears to be so homogeneous over such vast scales. How could the homogeneity of the universe be explained within a physical model? Well, in general, objects become homogeneous by allowing their different parts to come into contact and thus acquire common features. For instance, milky coffee becomes homogeneous by stirring it—allowing the milk to spread all through the coffee.

But the horizon effect prohibits precisely such a process. It tells us that the vast regions of the universe that we observe to be so homogeneous could not have been aware of each other at first. They could not have homogenized by coming into contact because they were unknown to one another. Within the Big Bang model, the homogeneity of the universe cannot be explained. Indeed, it looks rather spooky, almost as if telepathic communication occurred between disconnected regions.

Somehow, something must have opened up the horizons of the baby universe and produced its homogeneity, effectively producing a Big Bang model. At once we find that one of the riddles of the Big Bang universe, the riddle of its homogeneity in the face of the horizon effect, is prompting us to replace the early stages of the Big Bang theory with something more fundamental. The doors are open to speculation.

Above all, this was the riddle that was upsetting me in the winter of 1995, while I walked across the fields behind St. John's College. It looks easy to solve until you actually try. Then it becomes the ultimate nightmare, as I was finding out. But a second, even more tantalizing riddle was starting to bother me, the so-called flatness problem. It concerns the rather whimsical dynamics of expansion and how that relates to the "shape" of the universe, and I'm afraid it takes a bit longer to explain.

The story goes back to the time when Einstein felt that the universe should be static, before Hubble announced his findings. While Einstein refused to let go of his prejudices, the Russian physicist Alexander Friedmann worked out all the mathematics showing why and how the universe should be expanding as dictated by the general theory of relativity. Friedmann: the "underdog" who made the universe expand.

THE SPEED WITH WHICH Russian scientists jump into claiming priority over any discovery made in the West has become a running joke at international conferences. Give a presentation on toilet flushers and rest assured that some Dimitri or other will start shouting from the back row that the toilet with all its accessories was invented in Russia decades before the West even knew there was shit.

Sometimes, though, the Russians are actually correct, and modern cosmology is one such instance. The West seems to want to ignore the fact that after Einstein blundered in the late 1910s and before Hubble discovered the cosmic expansion in the late 1920s, Alexander Friedmann disclosed the intricacies of the cosmic expansion as predicted by the general theory of relativity. As his Russian colleagues point out, Friedmann should be put on the same footing as Copernicus, the man who placed the Sun at the center of the solar system, for he is behind a similar shift in cosmic perception, one that allowed for a nonpermanent universe.

Friedmann would perhaps be better known if he had lived a less eventful life, and if his unquestionable talent had been more conventionally applied. But his life fell prey to history, spanning the 1905 political unrest, World War I, and the Communist revolution and ensuing civil war. Writing to a friend in 1915 (while, elsewhere and better fed, Einstein finalized general relativity), Friedmann satirized his ordeal in the following way: "My life is fairly even, except for such accidents as a shrapnel explosion twenty feet away, the explosion of an Austrian bomb within half a foot, which turned out almost happily, and falling down on my face and head, which resulted in a ruptured upper lip and headaches. But one gets used to all of this, of course, particularly seeing things all around which are a thousand times more awful."

Friedmann's mathematical ability was supreme and it shone through even in those messy times, typically in matters such as the calculation of trajectories of bombs falling from planes. He would often double up as the aeronautical scientist and the test pilot.

These experiences embittered him considerably, and one gets the impression that his self-effacing personality resulted partly from a sense of shame for all the horrors with which history had blended his science. But on the rare occasions when things quietened down, he produced striking, cutting-edge research with more peaceful applications, in fields as disparate as meteorology, fluid dynamics, mechanics, and aeronautics, to mention but a few. He also became a pioneer balloonist, breaking altitude records while performing innovative meteorological and medical experiments aboard.

His energy was electrifying and unique. At his quietest, his activities combined a heavy load of teaching, administration, and research. As an administrator, he was instrumental in the creation of many new Soviet research institutes, and he was always busy collecting funds for salaries, laboratories, and libraries. As a teacher, he would often take on a minimum of three full-time jobs simultaneously.

In 1922, at the age of thirty-four, Friedmann turned his wide-ranging interests to the theory of relativity, studying with great application the general theory. Due to the war, and later to the blockade of the Soviet Union, general relativity arrived in Russia with several years' delay. Friedmann was one of the first in his country to study the new theory and to write in Russian about it, preparing various textbooks and popular science monographs on the new theory, eager to ensure that the new generation would not miss out on the crucial developments taking place. On the side, he also started doing his own little sums with the theory, playing with the new toy Einstein had presented to physicists.

Very little is known about Friedmann's personality; he is one of those people who are better known for their actions. So although we can't pretend to understand his reasoning, it is uncontroversial that Friedmann did not share Einstein's cosmological prejudices. When he applied the equations of general relativity to the universe as a whole and obtained an expanding universe, he did not run away in panic. He just took it at face value—no cosmological constant fudging—and in 1922 published his findings in a German journal. Ahead of Hubble's observations he thus *predicted* the expanding universe.

Friedmann's publication annoyed Einstein intensely, leading to a further twist in the story, which would rub the cosmological constant thorn even deeper into Einstein's flesh. At the time, Einstein hoped that his field equation would have one cosmological solution only— his static universe, which could thus be proclaimed the automatic winner purely on theoretical grounds, dispensing with troublesome astronomical observations. He believed that although there might be other solutions, these would be proved inconsistent with his field equation for one reason or another. Hence Einstein, upon reading Friedmann's paper, felt that the results reported not only had nothing to do with the real world but also were mathematically incorrect.

Therefore, somewhat against his character, Einstein published a nasty note in the same journal, a few weeks later, attacking Friedmann's work. He wrote: "The results concerning the non-stationary world contained in [Friedmann's] work appear to me suspicious. In fact it turns out that the solution given does not satisfy the field equations."

There is little doubt that Friedmann, like everyone else, revered Einstein, and he was likely very distressed upon reading Einstein's note. It must have been with a sense of doom that he carefully repeated his calculations over and over again. Had he really made such a stupid mistake? Finally, Friedmann had to accept the unbelievable: The great Einstein had made a mistake and Friedmann's original calculations were right. He wrote a respectful letter to Einstein clarifying his calculation, explaining where he felt Einstein had erred. It was such a simple algebraic error that Einstein saw at once that he had got his own sums wrong, and—presumably with some embarrassment—retracted his earlier note. He must have been very disappointed, not so much because he had made a mistake, but because his equation did not provide a unique solution to the universe in accordance with his most cherished beliefs.

In his retraction note, Einstein conceded gracefully: "In my previous note I criticized [Friedmann's work]. However my criticism . . . was based on an error in my calculations. I consider that Mr. Friedmann's results are correct and shed new light. They show that in addition to the static solution there are time varying solutions." Nonetheless, the handwritten draft of this note has survived, and in it one can read the crossed-out phrase: "A physical significance can hardly be ascribed to these solutions."

It is obvious that Einstein would have loved to add that last phrase to his note. But he knew that the evidence for such a statement was nil, and his honesty prevailed.

FRIEDMANN'S ARTICLES are an outstanding piece of research, and I must describe them in some detail, for they define the basic model of the universe upon which every other cosmologist, myself included, builds. They are also the basis for the nasty riddle of flatness. He introduces three types of models: closed or spherical models, open or pseudospherical spaces, and "flat" universes. These terms describe the shape of space, the basic fabric of the universe. He then shows that according to general relativity—at least if one does not play dirty tricks with Lambda—these models *must* be expanding, thus truly *predicting* Hubble's findings.

Without these papers, Hubble's discoveries would make little sense. It is sometimes said that we should never believe a scientific theory until it is verified by experiment. But a famous astronomer has also stated that we should never believe an observation until it is confirmed by a theory. Friedmann's papers, some ten years ahead of Hubble's findings, provided just such a theory.

Friedmann starts by clarifying the notion of cosmic expansion, establishing the interpretation we use today, and removing a few paradoxes that may crop into the theory otherwise. He shows that this expansion is a geometrical effect rather than the mechanical motion everyone pictures it to be. I must admit that I have played along with this erroneous interpretation so far, but let me now make amends and explain more carefully what expansion actually means according to the theory of relativity.

In the relativistic picture of expansion, the components of the cosmological fluid, that is, the galaxies, are encrusted in space and therefore are not moving relative to space; however, space itself is in motion, expanding and creating more and more room between any two given points as time goes by. Thus, the distance between any two galaxies increases in time, creating the illusion of mechanical motion. But in reality, galaxies just sit there, contemplating the spectacle of the universe creating more and more space in between them. This may seem a subtle point, but try to digest it. It is the source of many a misunderstanding in cosmology.

A possible analogy is to think of a hypothetical Earth, with its inhabitants confined to its surface, unable to view it from space. But now consider such a surface expanding outwards, like a swelling Earth, even though outer space is still inaccessible to its prisoners. Looking at the cities on the surface of such an expanding Earth, we would find that indeed they are not moving; however, the distance between them is increasing. Cities don't have legs to move about with, but somehow the dynamics of the space they inhabit creates an illusion of motion because their separations are changing.

This fine point is essential for the self-consistency of the theory. If the cosmic expansion were a real motion, we could easily run into paradoxes. For instance, Hubble's law states that the galaxies' recession speed is proportional to their distance. If such a speed were a genuine speed, describing a standard motion in a fixed Newtonian space, then we should be able to find a distance beyond which that recession would occur faster than the speed of light.

In fact, the speed of all galaxies with respect to the space that contains them is zero, in close analogy to those imaginary cities living on an expanding Earth. However, the distance between galaxies increases in time at a rate possibly faster than the speed of light if you consider suitably distant galaxies. There is no contradiction between the two statements just made, and so there is also no paradox or conflict with the special theory of relativity.

Nonetheless, Hubble's law can still be understood in Friedmann's picture. Friedmann's idea is that we live in an expanding universe, modeled as a space where all distances are multiplied by a number he called the *expansion factor* or the *scale factor*. This number keeps increasing with time, thereby describing the geometrical expansion. But because all distances are multiplied by this factor, the larger the distance the greater the increase. This would not happen if we were to add a number to all distances; but instead the expansion factor *multiplies* every distance, so that the larger the distance the more it increases as time goes by.

As a result, if we revert to Hubble's picture and describe expansion as a real motion in a fixed background space, it looks as if the "velocity" is proportional to the distance, that is, Hubble's law. But Friedmann's picture is far more sophisticated. It shows that the outward motion Hubble eventually observed does not actually have a center. Any observer has the illusion that he is the center of an outwards rush, satisfying Hubble's law, because in reality the whole of space is being stretched, at the same rate everywhere.

HAVING MADE THIS important point, Friedmann, like Einstein, postulated that the cosmological fluid is homogenous, that it looks the same, or has the same properties, everywhere. This was done out of intuition and mathematical convenience (not to mention laziness), rather than data. Remember that all these developments took place years before Hubble's discoveries. Indeed, to be historically more precise, the perspective from which Einstein and Friedmann worked was that of a homogeneous fluid of stars rather than of galaxies, which they knew nothing about. But by a miracle they both chose the right assumption, albeit with the wrong ingredients.

Assuming homogeneity drastically limits the number of possible geometries or space-times that can be used to describe the shape of the universe. Indeed, if matter produces curvature, and if the cosmological fluid density is the same everywhere, then the curvature of the universe should be the same everywhere. This rules out irregular, wildly curved shapes—for instance, the universe cannot be shaped like an elephant, a beast that is far from homogeneous. In fact we are left with only three distinct possibilities.

The simplest one is that of a three-dimensional space having no curvature at all—that is, Euclidean space. To help us visualize the other cases, in Figure 5.1.a I have plotted a two-dimensional analogue of the three-dimensional real thing, a flat infinite sheet. It may surprise you that this flat surface is a possibility at all, given that matter produces curvature; but remember that what matter actually curves is space-time, and we have not yet described the time side of the story for this universe.

To do so, Friedmann stated that *all* distances on this sheet are to be multiplied by the scale or expanding factor of the universe. This factor may possibly vary in time, thereby describing the temporal dynamics of such a universe. It is the whole thing—flat surface plus

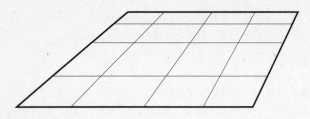

FIGURE 5.1.a

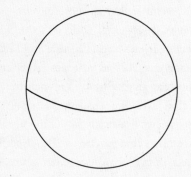

FIGURE 5.1.b

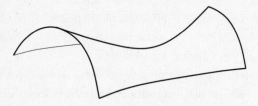

FIGURE 5.1.c

(a) A flat surface. (b) A sphere. (c) A portion of a hypersphere. The hypersphere is actually infinite and looks like a saddle at every point.

its time-dependent scale factor—that defines the proper space-time model, and that should be curved by matter according to Einstein's field equations. And, indeed, when Friedmann plugged this geometry into the field equation, he found that the scale factor is curved. In Figure 5.2, I have plotted the scale factor for this universe. You see

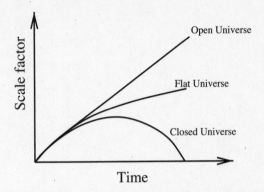

FIGURE 5.2 The evolution of the scale factor in a flat, closed, and open universe. Closed models expand to a maximum size, then recontract and die in a Big Crunch. Open models eventually start expanding without deceleration, as if gravity has switched off—they ultimately escape their own gravity and become empty worlds. Only flat models avoid these extreme fates.

that it increases in time but that the rate of increase decelerates. This deceleration may be interpreted as the curvature of space-time. The fate of this universe is unique, as we shall shortly find out, and another glance at Figure 5.2 shows that it expands forever, gradually decelerating in this motion, but never quite stopping.

The other two homogeneous spaces are more complicated. One is the sphere,* which also has the same curvature everywhere. That's easy enough to visualize, but remember that we are talking about a three-dimensional sphere, not its two-dimensional analogue. I have plotted the Mickey Mouse version in Figure 5.1.b: If you can visualize the real thing, good for you—I can't. This has never stopped me from working with three-dimensional spheres, but that is the advantage of mathematics. It enables us to play with things our brain cannot cope with.

If you found 3-D spheres baffling, well ... the third type of homogeneous space is even worse. It is called a pseudosphere, or,

---

*When scientists refer to spheres they mean the surface of the sphere only, which of course is two-dimensional for standard spheres.

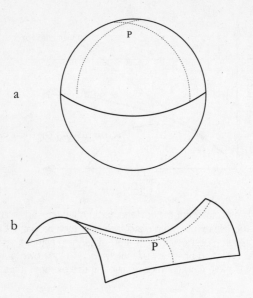

FIGURE 5.3 (a) Cutting two orthogonal sections through any point (p) in a sphere produces two circles curving in the same direction (here both circles curve downwards). (b) Performing the same operation on a hypersphere leads to circles curving in opposite directions (here one curves upwards, the other downwards).

sometimes, an open universe. A piece of its two-dimensional analogue is plotted in Figure 5.1.c, but the whole thing is actually infinite and looks like the saddle of a never-ending horse. To help you explore the meaning of a pseudosphere, I have played a trick in Figure 5.3: I have cut sections of spheres and pseudospheres along two orthogonal directions. In a sphere, you get a circle in either direction; for this reason we sometimes say (incorrectly) that a sphere is the product of two circles. A similar thing happens with the pseudosphere, but the two lines now curve in opposite directions. Hence we say that the pseudosphere has negative curvature, whereas the usual sphere has positive curvature. Depending on whether or not we add this twist to the product of the two circles, the space is either finite (like a sphere) or infinite (like a pseudosphere).

To describe how these last two spatial surfaces are to be combined with time to produce a space-time, we should also multiply all the distances living on them by a scale or expansion factor, which can depend on time. But when Friedmann plugged these geometries into Einstein's field equation, and examined the history of their expansion factor, he found that these spaces have very unsavory fates in contrast to the flat model considered above. He found that the spherical universe expands out of a Big Bang, but eventually comes to a halt, starts to contract, and then dies in a Big Crunch. The pseudospherical universe, on the contrary, expands out of a Big Bang and never stops expanding. However, unlike the flat model, its expansion does not decelerate forever, but eventually finds a steady rate. The time evolution of the scale factor as obtained by Friedmann for the three possible models is plotted in Figure 5.2.

We have seen this antithesis before. It reflects nothing but a tension we have already discussed: a war between expansion and the attraction of gravity, or the outwards swell of space versus the force of gravity pulling everything back together. A closed or spherical model is tuned so that gravity eventually gets the upper hand over expansion. Expansion proceeds, always being decelerated by gravity, until it is finally stopped, precipitating the universe into faster and faster collapse and contraction, towards the abyss of the final crunch. An open or pseudospherical model is one in which the war is won by expansion, when finally the universe escapes its own gravity. For a while, gravity is strong enough to decelerate expansion; but in the end expansion is so fast, or, seen in another way, everything has been diluted so much, that gravity becomes irrelevant. For this reason expansion stops decelerating, starting an epoch in which the universe has "escaped itself," essentially becoming empty.

Treading a fine line between these two models, we encounter the flat model—a rather British compromise—where a perfect balance has been struck between the powers of expansion and gravity. Expansion never frees itself from gravity, but gravity never brings expansion to a halt and collapse. The universe expands forever with phlegmatic moderation, neither giving way to gravity and a catastrophic implosion, nor to hysterical expansion and emptiness, sensibly avoiding disaster or death to live into a late and venerable age.

The longevity peculiar to flat models is crucial. Only such a universe lives long enough for matter to clump together to form galaxies and stars, and for immense time scales to be available for the production of structures and life. We cannot hurry the slow process by which natural selection cooks up intelligence, and in only one class of models do we have the necessary time without the threat of a cosmic hecatomb.

The trouble is, flatness is inherently unstable. It relies on the precarious adjustment of the powers of cosmic motion and gravity, a miraculous avoidance of two universal cataclysms. The slightest deviation from flatness and space-time quickly closes in upon itself or becomes saddle-shaped and empty, in either case hurtling towards a vertiginous death. In effect, it appears that the universe must have been walking a tightrope for 15 billion years—something highly improbable if not downright impossible. This is known as the *flatness problem*—the second of the Big Bang riddles. It has harassed cosmologists ever since Friedmann unveiled the landscapes of relativistic cosmology.

A possible description of this arm wrestling is a number called Omega (after the Greek letter). Omega is roughly the ratio between the gravitational energy of the universe and the energy contained in its outward motion. A flat universe has equal amounts of both at all times, and so Omega is equal to one. A closed model has Omega larger than one, for its gravitational energy is larger than its kinetic energy; an open model has a value of Omega smaller than one.

An equivalent way of expressing Omega is to define, for a given expansion speed, the value of the matter density that produces an

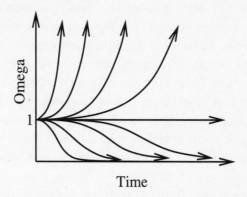

FIGURE 5.4   The instability of a universe with Omega equal to one. Any slight departures from a flat world quickly develop into larger and larger deviations.

amount of gravitational energy exactly balancing the expansion energy. Such a density is called the *critical density*, in a parody of nuclear weapons. This is the density needed to maintain Omega=1, or flatness. If the cosmic density surpasses the critical density, gravity eventually wins, and we live in a closed model. If we find ourselves in a universe with a density below the critical value, we know that eventually the universe will escape from itself into a gravity-free explosion, and we live in an open model. Not surprisingly, Omega may be written as the ratio between the actual cosmic density and the critical density, its value describing the current state of this gigantic tug-of-war.

What makes the flatness problem so intractable is that, as the universe expands, any deviations from Omega equal to one increase dramatically, a fact I have shown in Figure 5.4. For a flat model, Omega equals one forever, but if the slightest advantage of one type of energy over the other is registered, that is, if there is even the slightest deviation of the cosmic density from the critical density, then things can only get worse; and, in fact, very rapidly do get worse.

Alan Guth, the father of the inflationary universe, relates that in the months preceding his great discovery, he felt particularly struck by this problem. In his early thirties, at a crucial stage of his career, he couldn't have cared less about cosmology, which in those days was far from being a respectable branch of physics. Cosmology was seen as the sort of scientific endeavor a young man should avoid like the plague, leaving it to established older scientists suffering from rotting brains.*

More specifically, Guth was under pressure to publish "quick and dirty" in topics as mainstream as possible, but several accidents contrived to make him attend a lecture at Cornell in which the famous physicist Robert Dicke described the flatness problem.

Dicke shocked his audience by putting numbers into the problem. He showed that when the universe was one second old, the value of Omega must have been somewhere between 0.99999999999999999 and 1.00000000000000001. If Omega had deviated from one by more than this, either crunching or emptiness would have destroyed the universe a long time ago and we would not be here to discuss this outstanding philosophical matter. This remark impressed Guth so much that it derailed his career, precipitating the formulation of the theory of inflation. What could have tuned Omega so accurately?

I should inform you that Dicke did not choose the age of one second completely out of the blue. A crucial assumption in the calculations illustrating the instability of flatness is that the universe has been expanding, for otherwise Omega does not deviate from one. And Dicke knew well that we have observational evidence that the

---

*Curiously nowadays, when cosmology is no longer generally considered pestiferous, it is only the established old scientists who think that cosmology is a waste of time—a most peculiar social inversion.

universe has been expanding, in accordance with Friedmann's theory, since it was one second old.*

Before one second, we do not have any direct evidence of an expanding universe, only theoretical arguments. We *believe* that general relativity is still applicable before this time, in which case we can *infer* that the universe must have been expanding. We have no proof of this, but we also have no good reason to believe otherwise, and so we accept the extrapolation.

There is, however, a time in the past when we know that general relativity *must* break down. It's called the Planck time, and is very small indeed—0.(now write 42 zeros)1 second. We live in a quantum world, subject to random fluctuations. Unfortunately, we do not have a theory of quantum gravity, a theory that predicts how quantum fluctuations affect gravitational phenomena such as the motion of the Moon around Earth. But we can estimate the size of these fluctuations, and we find that they are invariably negligible in such problems as the trajectories of rockets or planets. So we don't have a theory of quantum gravity, but we don't need one, either.

The one tragic exception is the cosmological expansion before Planck time. At such early times, expansion, as predicted by the theory of relativity, is so fast that the quantum fluctuations cannot be neglected according to our best estimates of their size. Of course, we do not have direct access to this period in the life of the universe, so we cannot say with confidence that these fluctuations are important. But on the other hand we have no guarantee that we can rely on any results obtained without a full theory of quantum gravity. This argument will appear at several stages in this book: The so-called Planck

---

*This evidence could fill another book, but it's related to the way the universe cooks up elements heavier than hydrogen in a huge primeval H-bomb. The bomb would turn out to be a dud, and in contradiction with observations, unless the universe has been expanding, à la Friedmann, since the age of about one second.

epoch, preceding the Planck time, is a black box in the life of the universe as far as our theories are concerned. We cannot be sure about anything happening in this murky past.

In particular, we cannot be sure that the universe was expanding during the Planck epoch. Hence we only know that it has been walking the tightrope of flatness since then. But given that we have good reason to believe that the universe has been expanding ever since the Planck time, what values could Omega take back then for the universe to survive until now? The result is that it must be between 0.(can't be bothered to write 64 nines) and 1.(ditto—for 63 zeros)1, very close to 1 indeed.

I hope that by now I have explained myself better when I say that you have to contrive the Big Bang's initial state very carefully, turning its knobs to extremely carefully chosen numbers—and, moreover, do this by hand—to obtain something sensible. Why is Omega so close to one to begin with? Could it be exactly one? In either case, why? What mechanism contrives so as to produce such carefully chosen values of Omega and avoid catastrophe? The Big Bang model offers no answer to these questions. It merely gives you a range of possibilities, allowing you to select a universe with just the right value of Omega to obtain a model that describes incredibly well the world we live in. But you know that if you had chosen another model slightly to either side, you would have ended up with a complete monster.

And no theoretical principle assisted your choice, just the wish to fit the data. If you had chosen the initial value of Omega by *chance,* it would never have come up the way it did—that would be equivalent to winning the lottery some ten times in a row. Scientists started to feel that to some extent the Big Bang successes were the result of cheating.

Like the horizon problem, the flatness problem is begging for a speculation. Cosmologists need to start asking what actually happened during the bang, in that first instant of the universe's birth. What is hidden inside the Planck epoch, beyond the reach of relativity and Friedmann cosmology? Could this early life of the universe, its embryo stage, contain a specific process, some detailed hormonal chemistry that set these mysterious numbers to their unusual values? Why did we win the lottery so many times in a row?

BUT BEFORE LEAVING you to ruminate over these puzzles, let me add one more. The third riddle of the expanding universe is nothing less than that horrible devil unleashed by Einstein: the cosmological constant, or Lambda. It soiled Einstein's otherwise unblemished career, and he naturally repudiated it as soon as Hubble's discoveries were confirmed. After these tragic incidents, the cosmological constant fell into disrepute. It was perhaps the only serious error made by Einstein in his brilliant career. But once introduced, scientists could no longer justify why Lambda should be zero.

Recall that Lambda represents the energy of empty space, the gravitational power of the nothing—Einstein's special small door for his small cats. Einstein found that his theory of relativity allowed for a nonzero vacuum energy, as long as the vacuum was also very tense and gravitationally repulsive, and he used this fact to construct a static universe within his theory. To do so, he had to carefully tune the value of Lambda, so that its repulsive force exactly balanced the attraction of normal gravity. Hubble's discovery of the cosmic expansion killed the static universe, but not the cosmological constant. Indeed, as Friedmann already knew, only a contrived value of Lambda can make the universe static. A more general value of Lambda still leads to an expanding universe, so that Hubble's findings in no way rule out a cosmological constant.

But if the vacuum energy was not zero, how would it evolve in comparison to other forms of energy in the universe? Would it happily go away, as the universe expanded? Or would it dominate all the other species of the universe? And therein lies the third riddle of Big Bang cosmology.

FIGURE 5.5   A picture of the Coma cluster, a "pack" of galaxies.

The species of the universe go through something akin to natural selection. Some drop out, others dominate, leading to a scroll of epochs and ice ages not dissimilar to those experienced on Earth. So far, I have oversimplified the zoo that lives in our universe, so you may not be aware of what I mean by other species. I have so far mentioned only a cosmological fluid made up of galaxies because that is the most obvious component of our universe. But this is not the full story. Let me announce the other characters in the cosmic tragicomedy.

Figure 5.5 reveals a picture of the Coma cluster, a rich cluster containing over a thousand galaxies. In Figure 5.6 you see a picture of the same region in the sky but as seen with a telescope sensitive to X rays. X rays are the hallmark of very hot gas; in fact, gas heated up to millions of degrees. You see that the cluster is embedded in a soup of hot gas. It can be shown that most of the cluster mass is contained in this gaseous halo, proving that there is more, a lot more, than is apparent to the eye.

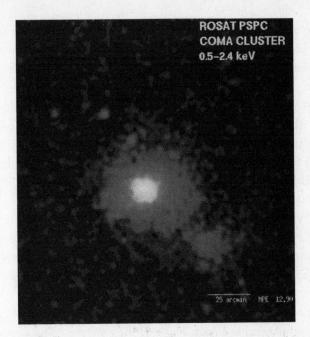

FIGURE 5.6 An X-ray picture of the same region of the sky as in Figure 5.5. The cluster is actually embedded in a giant cloud of very hot gas.

Similar experiments establish that what we can see with conventional telescopes is actually a very small fraction of the mass of the universe. Most blatantly we are surrounded by dark matter, stuff that does not shine, but that we can "see" through the unmistakable effects of its gravity. We can only feel its weight, and judging by the scales, dark matter constitutes most of the matter of the universe. We thus identify three species of matter in the universe: galaxies, hot gas, and dark matter.

But there is more. A further component is the so-called cosmic radiation, a sea of microwaves emanating from deep space, enveloping the vacuum in a lukewarm bath, warming everything by about 3 degrees. This background radiation was discovered by radio astronomers Penzias and Wilson in the 1960s, who initially mistook it for bird shit deposited in their antenna by a couple of pigeons. They cleaned and cleaned, swearing at those bloody pigeons who had decided to nest on their antenna. But no matter how obsessive they became, they could not scrub away the extra signal. Had the pigeons caused permanent damage to their precious instrument? In due time they realized they had measured something far more fundamental, the echo of another component in the universe, a cosmic fluid of radiation to be added to the matter fluid made up of galaxies, hot gas, and dark matter.

And these are the basic ingredients of the Big Bang as we know it.* The question is now: What happens to all this stuff as the universe expands? The answer is surprisingly simple—it only depends on whether or not a given species exerts pressure. We have seen before how pressure, or its negative counterpart, tension, may affect the gravitational power of an object. Indeed, give one species enough tension and it may even be gravitationally repulsive, as in the case of the cosmological constant. This was the trick played by Einstein to pull a static universe out of the reluctant hat of general relativity. We now find that pressure is also the determining factor in deciding the survival of the species as the universe expands. And this sets galaxies and dark matter apart in one corner, cosmic radiation in another . . . and, menacingly, the cosmological constant in yet another.

Let us start with the fluid of galaxies, which does not exert any pressure. Indeed, pressure is due to molecular random motions. Atmospheric pressure results from fast molecular motions, leading to a force acting upon any surface as the molecules bounce off it— that's what pressure is. But galaxies lack these motions, or they are negligible. Essentially, because they just sit there, they don't exert any pressure. Rather poetically, cosmologists call this pressureless fluid

---

*According to most theories, there should also be a sea of neutrinos, a cosmic neutrino background. This does not alter the argument in this book.

"cosmic dust," because dust also just sits around producing no pressure whatsoever.

As the universe expands, these pressureless galaxies are just stretched apart, or, more correctly, they remain encrusted in a space that expands to create more and more room in between them. If we could paint a given area of the universe red, then the red spot would spread with expansion, but the number of galaxies contained within it would remain the same. This defines the dilution rate for any dust-like or pressureless fluid subject to expansion: It just gets diluted at the same rate that the volume increases as a result of expansion. As far as we know, dark matter is also well modeled by a dust fluid, so its evolution with the expansion of the universe is the same as that of galaxies—it, too, gets diluted at the same rate any given volume increases.

Radiation is different in the sense that it does not just sit there. It is made up of photons, or particles of light, which naturally enough move at the speed of light, the maximum possible speed. For this reason a fluid of radiation, such as the cosmic microwave background, exerts quite a large pressure. How does this affect the evolution of this fluid when subject to the cosmic expansion?

As the universe expands, photons become more spaced out, but they also exert pressure on this expanding space. It is as if they were doing work, trying to help along the cosmic expansion, thereby using up some of their energy. As the universe expands, a given region painted red will expand and still contain the same number of photons. But each of these photons also gets weaker, exhausted from exerting a pressure force helping out expansion. A radiation fluid is therefore diluted by expansion at a faster rate than dust. It dilutes for two reasons: the volume expansion, plus the extra depletion resulting from the work done towards expansion by its photons.

This deduction has major implications for the history of the universe. If radiation is diluted faster than matter, then reciprocally the early universe must have been dominated by very dense, very hot radiation. Indeed, if one species is diluted faster than another, then it must always vanish at late times, but dominate in the very early universe. In other words, cosmic radiation must be the dinosaur of the universe—nowadays essentially extinct, but ages ago dominating the whole universe. The discovery of cosmic radiation therefore led to a particular version of the Big Bang model called the Hot Big Bang model: an expanding universe with a scorching past dominated by very energetic photons making up a sea of extremely hot radiation.

All this concerns the standard ingredients of the universe, what has now become the vanilla-flavored version of Big Bang cosmology. But what happens to this survival of the fittest if we add a cosmological constant? What is the fate of this hypothetical beast under expansion?

Recall that the vacuum energy is extremely tense, so that it must work against the expansion, resisting being stretched any further by expansion. But this means that, in perfect opposition with what happens for radiation, the cosmic expansion must transfer energy into the cosmological constant as it stretches the Lambda rubber band, forcing it to accumulate more and more tension. Hence expansion has a double effect upon Lambda: It dilutes its energy, but it also works against its tension, thereby transferring energy into it. These two opposing effects, the dilution due to volume expansion and the energizing effect of the tension, lead to a peculiar result. The energy density in the cosmological constant remains the same at all times, unaffected by the expansion of the universe. Expand the cosmological constant and its energy density stays the same!

This has a very dramatic implication. If at any given time in the life of the universe there is the slightest trace of vacuum energy, then as the universe expands and cosmic dust and radiation are diluted, Lambda will come to dominate the universe. Vacuum domination would mean a disaster: a universe very unlike our own. The skies would be black, ours would be the only galaxy in the sky, no cosmic radiation would be seen. How is it that vacuum domination has not occurred?

To put numbers into the puzzle, as before, let us first look at the one-second-old universe. It can be shown that the percentage of vacuum energy in the universe has to be smaller than 0.(add 34 zeros)1 percent in order to have avoided Lambda domination long before now. If we are more demanding, and assume that the universe has been expanding since the Planck time, then the initial contribution from the vacuum energy must have been smaller than 0.(write 120 zeros)1 in order to prevent vacuum domination before today.

We have another, even thinner, tightrope to walk.

THE RIDDLES OF THE BIG BANG are extremely annoying, and as early as the 1960s, cosmologists struggled to come up with solutions. But invariably they were found to be flawed. Perhaps the most entertaining early attempt was put forward by Yakov Zeldovich, a Russian cosmologist whose biography in some ways parallels that of Friedmann. His full formal education consisted of a total of six years of secondary school, a fact that probably explains his outstanding imagination and creativity. He was mostly self-taught—and even though he did not attend university, at the age of twenty-two he was awarded a Ph.D.

Zeldovich, like Friedmann, lacked ballast.* He worked on so many innovative ideas in cosmology that people sometimes think there were actually several cosmologists all with the same surname. Indeed, to distinguish the various formulae named after him, one usually hyphenates with his name the name of whichever Western scientist

---

*I apologize for the pun: Friedmann died prematurely from the effects of an early balloon ascent into the stratosphere, in which an altitude record was broken. Reading the pilot's and Friedmann's reports, one gains great insight into the Soviet space program—a series of successes always hovering dangerously on the edge of disaster, an unholy mixture of primitive, sometimes artisanal technology, and the infinite Russian ability to suffer.

rediscovered it several years later. It was Zeldovich who proposed the *bouncing universe* as a solution to the Big Bang riddles.

Here's the recipe. Take a spherical or closed model and let it expand out of the Big Bang. We know that such a model eventually turns around and implodes. According to general relativity, its final state is the Big Crunch, the ultimate collapse. But we know that as the universe plunges into the crunch, it eventually reaches contraction speeds that exactly mirror the expansion speeds achieved during the Planck epoch. In fact, the universe must enter another Planck-type era, a period in which unknown quantum gravitational effects come into play. The only difference is that now the universe is contracting rather than expanding. Cosmologists wondered whether these quantum gravity effects might turn the crunch around into a new Big Bang; that is, whether there might be a cosmic bounce.

The bouncing universe is sometimes also called the phoenix universe, because it goes from near-crunch to new Big Bang in an infinite cycle. Zeldovich was able to prove that each new cycle must be bigger (or last longer) than the previous one (see Figure 5.7).* He then tried to solve the Big Bang riddles using this feature.

At first glance this approach looked promising. However, after pages and pages of algebra, Zeldovich conceded defeat: The bouncing universe is actually not a solution to the Big Bang problems. If anything, it only makes these problems worse.

The riddles of the Big Bang are fascinating and dangerous. They hint at so much new physics, they scream for new cosmology, and yet they reveal nothing about their possible solution. They make it very easy for intelligent people to look totally idiotic. I recall a U.K. itinerant cosmology meeting in which Neil Turok, then one of the main opponents of inflation, got into a heated argument with someone

---

*The proof makes use of the second law of thermodynamics: entropy always increases. The maximum size of each cycle in a bouncing universe is related to its total entropy.

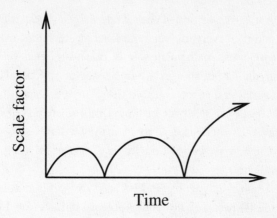

FIGURE 5.7   The scale factor in a bouncing universe. Every time the universe falls towards a Big Crunch it bounces back into a new Big Bang. The cycles get bigger, they last longer, and the universe expands to a larger maximum size after each "born again" experience.

who insisted that inflation was the only known solution to the horizon and flatness problems. Neil has a big mouth, and promptly retorted that it was not true, that he himself could come up with thousands of alternative explanations for the Big Bang riddles. Suppose that as the universe came into being, "something," some principle fell into action, which ensured that only a universe as symmetric as possible would be allowed. That would enforce a very homogeneous and flat universe, right? So there . . . that's a solution to the flatness and homogeneity riddles.

Hmm . . . I've always felt that copious use of the word "something" allows anyone to solve any problem, even insoluble ones. But there was a more obvious flaw. No sooner had Neil finished pronouncing his words than Mark Hindmarsh, sitting half asleep next to Neil, suddenly said, "Well, in that case, shouldn't the universe be Minkowski space-time?"

There was a one-second silence while people digested his comment . . . and then everyone burst out laughing, including me. If you know the relevant mathematics, it is obviously true: Minkowski space-time, the empty and gravity-free space of special relativity, is the most symmetric space available. It's so empty that it looks the same in every direction, space or time. Unfortunately, Neil's "something" principle would just lead to a blatantly wrong result—that we should live in a world without gravity—rather than solve any cosmological problem.

People laughed at what had indeed been a very misguided attempt to solve the flatness and horizon problems, but later on I thought that Neil had at least tried. And this is the hallmark of a good problem: frustratingly simple, and yet, as soon as you open your mouth and think you're going to solve it, absolute rubbish comes out no matter how clever you are.

However, Neil was wrong in one other respect. To be fair, and whatever people said at the time, by the mid-1990s there was only one answer to the riddles of the Big Bang—and that was Alan Guth's inflationary universe.

6   GOD ON AMPHETAMINE

IN THE LATE 1970s, cosmology was a bit of a joke. Particle physicists had made unprecedented progress in explaining the structure of
matter, isolating its fundamental particles as well as the fields mediating their interactions. Ever larger accelerators were being built; within
them physicists could concoct brutal particle collisions capable of
putting their theories to the test. These tremendous machines were
absorbing huge sums of public money, but everyone agreed that it
was money well spent because the results were very good indeed: The
theories were internally self-consistent (for the most part), and the
experiments carried out in accelerators corroborated the theories to a
ridiculous degree.

But whenever physicists tried to blend that enormous body of
knowledge, particle physics, with the Big Bang theory of the universe, nothing but sheer nonsense emerged. In principle, that combination should make sense, and even be a logical necessity, because the
very hot early universe should act like an extremely powerful, high-
energy accelerator. Therefore new particles should be produced in
the early universe in the same way they are produced in high-energy
collisions in accelerators. Reality, however, was a lot less clean.

Cosmologists were particularly interested in one type of particle,
the magnetic monopole, which had not yet been seen in accelerators
but which was predicted by key ideas that *had* been verified in accelerators. These monopoles should be produced in the very early uni-

verse, but how prolific would the early universe be in this regard? And would these monopoles decay away, once produced? If not, could there be relics, leftover monopoles still floating around us for the inquisitive scientist to detect?

The logic behind these questions has its roots in the discovery of cosmic rays in the 1930s. Cosmic rays consist mainly of particles produced within our galaxy, with energies far below those of magnetic monopoles.* Still, their energies by far exceeded the energy range of the accelerators in operation when cosmic rays were first discovered. In those days, Paul Dirac in Cambridge had just predicted the existence of antimatter, but its production was well beyond the capabilities of available accelerators. It was in cosmic rays that antimatter was first detected, many years before it could be produced on Earth.

The lesson was clear: Sometimes, particle physicists did not need high-energy accelerators to produce new particles; they had only to look up and the heavens would present them with a shower of high-energy particles, courtesy of the universe. Perhaps the same trick could be pushed to energies far higher than those of cosmic rays. Perhaps the early universe could be used as a very high-energy accelerator, capable of producing particles that we still could not produce on Earth, particles such as the magnetic monopole.

Still, the really big question was, How abundant should these relic monopoles be? And here starts the nightmare, because as soon as physicists put numbers into the problem, complete nonsense came out. The computed abundance of the relic monopoles left over from this hot early phase of the universe was so great that there should be nothing else in the universe except magnetic monopoles. Something had to be wrong, either with particle physics or with Big Bang cosmology.

Given the circumstances, scientists felt a bit at a loss. Here they had two very successful theories, particle physics and the Big Bang

---

*There are exceptions: the so-called ultra high-energy cosmic rays.

universe, both of which worked very well within their own domains. Scientists knew that logically these two theories had to overlap at some point, but wherever they did, the result was garbage. In the prevailing atmosphere of the 1970s, it is perhaps not surprising that cosmology took the full blame for the cataclysm. In those days, it was proclaimed that "cosmology is inconsistent with particle physics," the tacit implication being that no one should take cosmology seriously.

It looked as if the universe had been created by two gods on unfriendly terms.

IN THE LATE 1970s, the young Alan Guth was a mainstream particle physicist and therefore by all accounts should not have wasted his time with cosmology. But things were not going well for Guth. He had written several articles, yet his work was largely ignored. Nowadays, even Alan admits that his early articles border on the irrelevant.

He was then reaching that stage in a physicist's career when either he becomes a permanent fixture (in the academic jargon "gets tenure") or is summarily fired. This unforgiving duality hits most in their early thirties and is not widely known outside the physics community. But here are the facts: One beautiful morning the temporary contract market closes its doors to the aging physicist, and if you fail to get tenure at this stage you typically join the world of finance and feel frustrated about it for the rest of your life.

Given that Alan's papers thus far were less than successful, things did not bode well for him, and one can detect a tone of desperation lurking in his later accounts of these dark days. But people often do mad things when cornered, and at about this time Alan made a radical decision that would lead to the discovery of inflation: He would dedicate himself to "particle-cosmology," as the field would become known. He knew nothing of cosmology at the time, and he was moving into a field that physicists were then avoiding like the plague. Just to make matters worse, he started working on nothing less than the magnetic monopole problem.

Alan worked in collaboration with a colleague, Henry Tye, and they approached the problem in an unconventional fashion. They started to look for particle physics models that would *not* lead to a universe overstuffed with magnetic monopoles. This may sound innocent, but under closer inspection it is not. Their logic ran against the trends of the day: They were using cosmology to learn more about particle physics, as if cosmology were reliable enough for that purpose. A few centuries before and elsewhere, the Inquisition would have taken a keen interest in them.

To carry out their plan, they had to examine the details of the monopole production process very carefully. This entailed becoming experts in the field of the so-called phase transitions in particle physics, the process that creates magnetic monopoles in the early universe. You are certainly familiar with phase transitions in the context of water. It can be solid (ice), liquid (the stuff coming out of the tap), or vapor (loosely, steam). These three versions of water are more generally known as phases, and by changing the temperature you can unleash a phase transition. An example is the conversion of water into vapor, otherwise known as boiling, or of liquid water into ice, more commonly known as freezing.

Magnetic monopoles were produced in phase transitions affecting the stuff making up fundamental particles, but at melting temperatures, in degrees, of about 1 followed by 27 zeros. That such phase transitions should be present was part and parcel of the very successful particle physics theories of the day. On the other hand, you can't achieve such temperatures on your stove or even in the most powerful accelerators, so you might think that no one could ever defrost such stolid "ice." But as long as you considered its life sufficiently close to the Big Bang, the expanding universe might provide just the type of stove capable of generating such extreme conditions. The expanding, aging universe cools down, which means that reciprocally the early universe is very hot.

Alan and Henry, as well as others before them, found more precisely that the universe would be hotter than the required temperature for all times before 0.(now write 19 zeros and a one) seconds after the Big Bang. Therefore the usually "solid" particle stuff would be like a "liquid lava" during this period. As the universe expanded, and the temperature dropped, the primordial "particle liquid" would freeze into the rocky stuff making up the particles we know. Magnetic monopoles, in this analogy, are like tiny vaporous bags, fog if you like. They are a remnant of the hot phase, trapped inside minute cores. The trouble was that this primordial fog was more like an emulsion of supermassive cannon balls. How could we avoid a universe filled with a thick magma of superheavy monopoles?

After much hit-and-miss, Alan and Henry discovered a possible way out. They found that in some particle models the universe would "supercool." This term does not derive from the Californian vernacular, but merely means that as you drop the temperature of very pure liquid water, it may remain liquid well below freezing point. In fact it is even possible to supercool water to below minus 30 degrees centigrade. The supercooled liquid is extremely unstable, and the tiniest nudge causes an explosion of ice crystals. One may find supercool water and other fluids in nature. For instance, the blood of hibernating arctic squirrels may supercool to minus 3 degrees, when it would normally congeal. The supercooled blood still flows, since it remains a liquid, but the slightest disturbance will cause it to freeze, killing the squirrel; therefore, you should not disturb hibernating arctic squirrels.

A similar process may occur in particle physics, and Henry and Alan claimed, erroneously, that supercooling could eliminate the menace of monopole overpopulation.* They published a paper describing their discovery, and although this paper is essentially

---

*The idea is that fewer monopoles would be produced in a phase transition delayed by supercooling. Indeed, you should form roughly 1 monopole per horizon volume, and the later the transition occurs the larger the horizon. It turns out that this is not enough to avoid monopole domination.

wrong, its "side effects" sparked a revolution in cosmology. Indeed, just as they were about to submit their paper, two dramatic things happened to Alan that would lead to the serendipitous discovery of the inflationary universe.

First, Henry abandoned ship, leaving Alan to his own devices. Truth be said, you rarely know when you are up to something really important, but Henry was also under considerable pressure to stop working on all this nonsense. Alan relates that at about this time Henry had been told by a senior scientist that his work on monopoles was too "esoteric" to further the case for a promotion he was applying for. Henry then made a cardinal mistake: He listened to a senior scientist who held power over his career. As a rule of thumb, one should always assume that such people are senile; Henry therefore gave up the extraordinary work he was developing with Alan at its crucial stage.

No doubt Alan must have been under similar if not worse pressures. He was not just jeopardizing a promotion—he was one step away from the end of his scientific career. Left alone, nonetheless, he was foolish enough to proceed. There is a Portuguese saying: Lost for a hundred lost for a thousand. Alan's career was derailing so badly at this point that what the hell, he might just as well pursue this "esoteric" stuff all the way.

A major unexplored issue was the gravitational properties of the supercooled matter, a question that Henry had raised before he decamped. Alan now decided to work out what type of gravity would emanate from such an extreme form of matter.

It was at this point that Alan made an astonishing discovery. He found that the supercooled gluck of his particle theories was a very tense material that would be gravitationally repulsive, which indeed behaved just like a cosmological constant! Not quite like a proper Lambda, but like a temporary Lambda, one that would remain switched on only while the universe was supercooled.

Once again, Einstein's biggest blunder was back.

UNLIKE HENRY, Alan's instincts did not deceive him. He felt at once that his discovery had the unmistakable signs of a big break-through. He immediately became very excited about his brainchild, and the next day he rushed to tell a prominent colleague. Perhaps not surprisingly, his enthusiasm was met with a cold reaction, as well as the reply, "You know, Alan, the most amazing thing is that they pay us for this." Henry was not the only one to fail at first to see the extraordinary reach of the new idea.

Significantly, Alan ignored these comments and reactions. By this time, he had made an even more amazing discovery: The supercool universe, with its temporary cosmological constant, was a solution to nearly all the cosmological puzzles! At long last the two enemy gods—particle physics and cosmology—had fallen in love, and from their union it appeared that particle physics was the missing link needed to explain the outstanding mysteries of Big Bang cosmology.

The supercool universe is a brief affair with the cosmological constant, a temporary flirtation with Einstein's biggest blunder. This naughty episode in the life of the baby universe was dubbed by Alan *inflation.* The origins of this expression can be traced to the fact that the cosmological constant is gravitationally repulsive and causes the universe to expand extremely rapidly, its outwards rush accelerating into faster and faster expansion, rather than decelerating, as it would in the presence of normal, attractive gravity. Therefore the size of the universe (as well as all distances between objects partaking in the cosmic expansion) increases by a huge amount during this short episode in the life of the universe. Thus the term *inflation:* While the universe is dominated by supercooled matter, its size inflates.

Inflation is really like drugging the baby universe with speed. The supercool union of the hitherto unfriendly gods was blessed by amphetamine, and this made the universe *inflate* rather than just expand. The early orgy of expansion in the universe comes to an abrupt end as soon as the supercooled particle stuff finally freezes.

Suburban normality is then reestablished; the universe regains its Hot Big Bang tag and decelerated expansion resumes its normal course.

But this early affair with Einstein's biggest blunder has dramatic consequences for the later life of the universe. On that same long night when Alan discovered the inflationary universe, he found that under inflation the usual instabilities of the Big Bang model became stabilized. Flatness, rather than being an unlikely tightrope, became the inevitable valley into which the inflationary universe had to flow. Horizons would open up and bring the whole observable universe into contact, sewing together into a fine homogeneous whole what otherwise looked like an ungodly patchwork of disconnected islands. By the time the universe left its inflationary phase it would be sufficiently finely tuned to walk the tightrope without falling off. Inflation was a solution to the instabilities of Big Bang cosmology. The sphinx and her riddles were about to take a beating.

TO SEE HOW INFLATION solves the horizon problem, I have to start by admitting that I have so far simplified the problem. However, simplifications are often unavoidable if one wants to discuss physics without mathematics—and the version of the horizon problem I have given you is qualitatively correct for Big Bang models, and even for VSL. Nevertheless, it breaks down for inflationary expansion because a curious subtlety comes into play. It then becomes apparent that in the definition of horizon distance we have been ignoring the interplay between expansion and light's motion. Paying proper attention to this detail paves the way for the inflationary solution to the horizon problem.

Recall that the horizon problem stems from the fact that at any given time, light—and therefore any interaction—can only have traveled a finite distance since the Big Bang. The baby universe is therefore fragmented into horizons, or regions that cannot see one another. Such a patchwork of disconnected horizons is enormously irritating to cosmologists. It precludes a physical explanation, that is, one based on physical interactions, for questions such as why the early universe is so uniform.

We would like the cosmic uniformity to result from the whole universe coming into contact, its temperature thereby equilibrating throughout a homogeneous sea. But instead, the early universe is split into a multitude of regions out of touch with each other. Within the framework of standard Big Bang theory, uniformity can be achieved only by fine-tuning the initial state of the universe, that is, by carefully arranging for all these separate regions to be prepared with exactly the same properties to start with. This is very contrived and essentially not an explanation at all; it's more like an admission of defeat.

But what really is the size of the horizon? We have previously said that the horizon radius is the distance traveled by light since the Big Bang. Computed in the most straightforward manner this means that, for example, the horizon radius for the one-year-old universe is one light year—the distance traveled by light over one year. But is this exactly true?

The answer is no, due to the subtlety I have been warning you about. Travelling in an expanding universe entails a surprise: The distance from the departure point is larger than the distance traveled. The reason is that expansion keeps stretching the space already covered. By analogy, consider a driver who travels at 100 km per hour for one hour. The driver has covered 100 km, but if the road has elongated in the meantime, the distance from his point of departure is greater than 100 km.

Or imagine a cosmic motorway, realized if the Earth were expanding very fast. Then a trip from London to Durham might show on the odometer that 300 miles were traveled, whereas the actual distance between the two places by the end of the trip could be 900 miles.

Similarly, in a 15-billion-year-old universe, light would have traveled 15 billion light years since the Big Bang. However, the distance to its starting point would be roughly 45 billion light years. These are the numbers that actually emerge from a proper calculation, so that, because of this peculiar effect, the current size of the horizon is three times the naïve expectation.

This fact does not change the essence of the horizon effect in Big Bang models. Sure enough, the horizon is larger than the naïve expectation, but it can be shown that the horizon size still increases in time, which is the key feature behind the horizon problem. This means that the horizon still becomes tiny in the past when compared to its current size. And so we can still conclude that far-away objects are seen as they were long ago, when the horizon was much smaller, and therefore they may well be outside each other's horizons. The observed homogeneity of the past, faraway universe is therefore puzzling because its various regions are out of touch with each other, with or without this "triplication" effect.*

However this is true only in the context of normal, decelerated expansion. The whole argument breaks down under accelerated or inflationary expansion. With inflationary expansion, the distance traveled by light since the start of inflation essentially becomes infinite. Expansion is then so fast that its action of stretching the distance already covered works faster than the actual forward movement of light. For this reason, inflationary—or accelerated—expansion is sometimes dubbed superluminal expansion, which is not entirely correct but is certainly descriptive enough. The point is that under amphetaminic expansion, light travels a finite distance, but expansion works "faster than light," infinitely stretching the distance separating the light ray from its departure point.

---

*It is true that at the Big Bang moment itself, the universe is reduced to a point, but this does not mean that the universe is all in contact. Indeed, the horizon is also reduced to a point at the bang, and if you ask how many horizons fit in the universe at the creation moment, that number is infinitely large. Somehow at the bang, the horizon is a point infinitely smaller than the universe.

Therefore, inflation opens up the horizons. The whole observable universe nowadays was, before inflation, a tiny bit of the universe well in causal contact. Seemingly disjoint regions could then have communicated with one another and reached a common temperature in the same way that mixing cold and warm water soon leads to luke-warm water throughout. This small homogeneous patch was then blown up by a period of inflation into a vast region, much larger than the whole 45 billion light years we can see nowadays. The horizon problem is only present if we postulate standard Hot Big Bang expansion all the way back to time zero. If instead we introduce a brief period of inflationary expansion into the life of the very early universe, we solve the horizon problem.

THE FLATNESS PROBLEM would be the next victim of inflation.

We have seen that the cosmological constant has eccentric properties quite unlike any material we encounter in our everyday experience. For one thing, it is gravitationally repulsive, a most unusual feature. But remember it also displays another freaky type of behavior: Its energy density is not diluted by expansion but remains constant.

Normal forms of matter become more diluted if they are placed inside a box and then the box size is expanded, allowing its contents to spread out inside. This is as true for corn flakes as for the cosmic dust introduced earlier. If you double a 1-cubic-meter volume containing 1 kilogram of cosmic dust, its density is halved. You still have 1 kilogram of dust, but now that it occupies twice the volume, its density becomes 0.5 kilogram per cubic meter.

Not so with Lambda: In the same circumstances you would still have 1 kilogram per cubic meter of Lambda over your 2 cubic meters, so that somehow you have ended up with 2 kilograms of Lambda where at first there was only 1. A box of Lambda contains the same number of kilograms per cubic meter even as its volume is doubled and so contains twice the initial mass or energy.

This unusual feature, as we have seen, is due to the fact that Lambda is a very tense material, so that expansion energizes the Lambda sea in the same way that a rubber band accumulates energy when it is stretched. But whereas for a rubber band the effect is only a very small contribution to its internal energy, the Lambda tension is so high that the accumulation of tension energy acts to compensate the dilution associated with expansion. Expansion dilutes the Lambda energy, but then tension exactly makes up the deficit.

It is this disparity between the behavior of Lambda and ordinary matter that led us to the statement of the cosmological constant problem. The slightest trace of Lambda would soon lead to a universe with nothing but Lambda. Cosmic expansion would ensure that all normal matter would be diluted, whereas the Lambda density would remain constant. Soon there would be nothing in the universe but Lambda, which would rule the universe forever.

The Lambda problem is in some ways similar to the flatness problem. Both arise from domineering tendencies, of either curvature or of Lambda. Recall that the flatness problem is the instability of the flat Friedmann model. Homogeneous cosmological models can be flat, spherical, or open (also called pseudospherical). It was found that slightly spherical models would become violently more and more curved until they closed up on themselves in a tragic Big Crunch. Open models, on the other hand, would open up so much that they would end up in sterile emptiness, completely devoid of any matter. In either case, curvature likes to dominate matter, and the outcome is something very different from the universe we live in.

This is just to recap ideas that have already been presented. But now let me draw your attention to one crucial thing I have failed to discuss. The flatness and Lambda problems concern the antisocial tendencies of curvature and Lambda to dominate over ordinary matter in the course of cosmic expansion. But what about a tug-of-war between curvature and Lambda themselves? How might the cosmological constant and the flatness problems interact?

What Alan Guth found was that one villain would destroy the other: The domineering tendency of curvature was simply no match for the brutality of Lambda. He found that a flat universe is only unstable in a fight between curvature and ordinary matter. Against Lambda, curvature would hopelessly lose the battle, so that a *flat* (and also Lambda-dominated) universe would reign. It was as if curvature *did* get diluted by expansion, although less so than ordinary matter. Thus, curvature would dominate normal matter, but would in turn be dominated by something that was not diluted by expansion at all, such as supercooled matter, or a cosmological constant.

But inflation, unlike a proper Lambda, is not a true dictator. The inflationary Lambda is a parody of a real dictatorship and soon ends its rule of its own accord, when the supercooled matter finally "freezes." Inflation is a Lambda dictatorship in disguise—a temporary Lambda that decays into ordinary matter when its time is over. But while this semblance of dictatorship is in effect, a less potent tyrant, curvature, is violently suppressed. When this job is done, democracy is reestablished, the dictator turning into all the normal matter and radiation in the universe. This is the ingenious inflationary solution to the flatness problem.

If you like Omega, the ratio between gravitational and kinetic energies in the cosmic expansion, we may rephrase these findings by stating that Omega equal to 1 is no longer unstable under the rule of Lambda, but instead becomes what scientists have dubbed an attractor (see Figure 6.1). This is true whether we have a proper Lambda or only a temporary one (i.e., inflation). It depends only on the fact that Lambda has a very high tension and is therefore not diluted by expansion.

Alan then remembered that lecture by Dicke he had attended long ago. Dicke had stated that the survival until nowadays of the one-second-old universe required Omega to be tuned between 0.99999999999999999 and 1.00000000000000001. He quickly found that even a modest amount of inflation would dilute curvature so

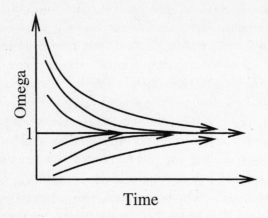

FIGURE 6.1   The Omega equals one universe becomes an attractor during inflation, and so inflation solves the flatness problem.

that the one-second-old universe would sport an Omega between 0.(write a few pages with 9s) and 1.(write another few with zeros and add a one at the end). Inflation was an extremely effective method for suppressing curvature and enforced the required tuning necessary for solving the flatness problem.

AT THE END OF INFLATION, the supercooled universe decays into the matter and radiation of a normal Hot Big Bang universe, the temporary cosmological constant switches off, and crazy amphetaminic expansion gives way to the standard decelerated expansion peculiar to attractive gravity. The normal course of the Big Bang resumes, but its worst nightmares have been staved off. It is no longer a coincidence that the universe is homogeneous across so many disconnected horizons. All these separate horizons went to the same nursery school. The instabilities of the sensible brand of Big Bang models (the flat ones) are no longer a concern. A period of inflation finely tuned the universe. It gave it the stability at birth required for the universe to cope with its "instabilities" in later life.

The only problem inflation does not solve is of course the Lambda problem itself—to some extent inflation is built upon it. If in addition to the temporary Lambda provided by supercooled matter there is also a permanent cosmological constant, the latter will not be suppressed by inflation. The energy densities both of fake and proper Lambdas remain constant during inflation, and therefore at a fixed ratio. Hence a genuine Lambda would still threaten to dominate the universe any time after inflation.

But still the battle was won on all other fronts. The daring strategy was to use one of the Big Bang riddles to solve the others, in a sense turning the sphinx against itself. The sphinx was not fully defeated, but it was very badly hurt and left with but a single weapon. Such was the remarkable achievement of the inflationary theory of the universe.

To close the story of inflation, let me finally add that in its details Alan Guth's supercooled universe turned out to be a mere handmaid for the true inflationary paradigm. For various technical reasons, Guth's initial proposal was terminally flawed, but who cares—he had the main idea even if it was not in its final incarnation. It is sad that all too often credit is given not to the people who conceived a new theory but to those who come afterwards and clean up the fine details. Lee Smolin expressed this dichotomy in terms of "pioneers and farmers," the farmers frequently taking all the credit for the discovery of new territory. This unfortunate tendency did not prevail with inflation, and the man who charted the new territory received the full credit he deserved.

But to be fair, in inflation's case, those who came after Alan did a lot more than clean up the dirty floor. It took physicists several years of hard work to fix the flaws in Alan's initial proposal, and what came out in the end had many qualitative novelties when compared to Alan's initial proposal. The physicists who fixed inflation's teething problems were Paul Steinhardt and none other than my collaborator-to-be Andy Albrecht.* Andy was at the time a mere graduate student, and, in the great tradition assisting the history of science, Alan Guth's excellent book *The Inflationary Universe* is embellished with like-nesses of all the scientists involved in the creation of inflation, except the junior guy—Andy.

Nowadays, inflationary models have replaced supercooling with other, more effective mechanisms for producing inflation. They usu-ally introduce a special field, the *inflaton,* capable of driving a period of cosmic inflation and solving all the cosmological problems (with the exception of the Lambda problem) without running into the grief that Alan's first model had to face. Sadly, no one has ever seen an inflaton.

To put the record straight, I should also add that the monopole problem, which initially worried Alan Guth so much, is no longer regarded as a real cosmological problem. Again, it acted as a hand-maid for higher ideas, but ironically the early monopole pathologies are now blamed on particle physics models rather than on cosmol-ogy. And to be frank, perhaps one day the cosmological riddles them-selves will also be perceived as mere handmaids. They stimulated sci-entists' minds, but the theories of the early universe they conceived to solve these riddles vastly transcended their initial motivation. This is certainly true of inflation—but that is another story, requiring a whole new book.

I will conclude by stating the obvious: Alan never became a physics reject, working out derivatives for a price. After some initial healthy skepticism, scientists were quick to realize inflation's poten-tial. Indeed, inflation was such an overnight success that long before Alan's paper appeared in the scientific press most of the top American universities were fighting each other to enlist him into the

---

*In addition, there was a third physicist, but he gets so ballistic when he is not cited by name that I can't help not citing him here.

ranks of their permanent staff. By now you probably have noted that I have somewhat anarchist tendencies, or at least resent the strait-jacket of the establishment into which we are often forced to stuff our creativity. But I am not religious about it. Senior people some-times (quite accidentally) do the right thing, and the success of Alan Guth's career after he followed the lunatic path of inflation is testi-mony to this.

As the years went by, inflation's popularity among physicists con-tinued to grow. Eventually, inflation itself became the establishment. So much so that it gradually became the only socially acceptable way to do cosmology—attempts to circumvent it usually being dismissed as cranky and deranged.

But not on the shores of Her Majesty, Queen Elizabeth II.

| | | | | | | | | | | | | | | | | | | | | | | | | | | | | | | | | | | | | | | | | | | | | | | | | | | | |

PART 2

LIGHT YEARS

7 ON A DAMP WINTER MORNING

SOME ONE HUNDRED KILOMETERS north of London lies a flat expanse of lowlands, former marshes drained in modern times. They're swept by chilly winds and immersed in permanent gray, ensuring a miserable existence for the dwellers of their scattered hamlets and villages. Considering their vicinity to bubbling London, these so-called Fenlands provide a surprisingly rural setting, and green field after green field is populated more by cows than by people. In the middle of this bleak landscape is a quaint medieval town, its church and college spires rising up to the skies; it's been a widely famous seat of learning since the Middle Ages, when Oxford dons, running away from irate peasants, settled in what they hoped would be quieter surroundings, more conducive to intellectual pursuits.

Ever since then, the place has accommodated people with the necessary level of imbalance required to come up with new ideas. And this is the place to which I moved to study theoretical physics in October 1989. I was attracted by Cambridge's scientific reputation, its roots going far back to Newton, a reputation reflecting the bias towards natural sciences that had won it the nickname of Fenland Polytechnic.

At once I had mixed feelings about the place; but within the mixture I could identify an unmistakable pressure to come up with something different, something new. I find it hard to convey the confusion of good and bad energy I drew from the place, but I'll try.

On the positive side, I loved Cambridge's tolerance towards difference, and the way it promoted original thought. It's not just that you are sitting perhaps in the very same chairs that such famous physicists as Paul Dirac and Abdus Salam sat in. Or that the sink-or-swim mentality of the place builds such enormous self-confidence in those who do learn how to swim. Or that British manners will often excuse bad behavior so that everything is more or less tolerated (I once ended a disastrous evening by throwing up perilously close to the Master's wife; the following day, everyone behaved as if nothing had happened). It's not even that most dons have reached that stage of happy senility that unavoidably leads to hilariously eccentric behavior. It's all of these, and too many more to list here; but the overall feeling is that of a benign madhouse, inside which you feel that you have not earned your place unless you come up with at least one crazy idea in total disagreement with everything else proposed before.

That is the positive side of Cambridge, which will remain in my mind as the best part of my years as a research fellow at St. John's College.* But there is another side to the Cambridge experience that I found far less attractive. Cambridge is a place where the college fellows dine on a "high table," physically elevated above students' tables. It is a place where a surprisingly large number of people have been admitted at one time or another to a psychiatric hospital; I recall a tea party where nearly all mental disorders were represented. It is unfriendly to women and to foreigners; as a foreigner, I started liking the place only when I grew confident enough to return the compliments of xenophobia in kind. It carries forward the worst of the

---

*To those unfamiliar with Cambridge, it may come as a surprise to learn that the university itself provides students only with lectures and exams, with their real life taking place in some thirty affiliated colleges, where students receive tutorials, have rooms, and take their meals. Each college is ruled by a *Master* and an "upper tribe" of scholars called *fellows* or *dons*. The older colleges are like medieval fortresses, open to the outside only via a set of imposing gates, guarded by an army of overworked, amazingly rude *porters*.

class-bound British past. And of the British colonial heritage, with all
its pathetic chauvinism.

A simple story sums up in my mind this unholy mixture of humor
and creativity on one side and snobbery on the other. I did not actually witness this incident, which for all I know may be a "rural" myth,
but it certainly captures a great deal of the atmosphere I am trying to
describe. It appears that a student one night became inordinately
drunk, climbed up to a college roof (which is in itself a highly popular sport), and relieved himself on a passing porter. On being given
chase by the porter, the student committed the further heresy of
walking on the grass, a privilege that, like eating at the high table, is
reserved for college fellows. For these transgressions the student was
admonished by his tutor and fined: Twenty pounds for walking on
the grass, ten pounds for urinating on a porter.*

If this is a myth, it is not the only one of its kind. There is a great
profusion of similar stories, always obnoxious in a childlike fashion
and full of old-world ridicule. It is telling that some of these incidents originate from oddities in the university and colleges' statutes,
which were written in stone several centuries ago, so that they're
now a parade of anachronisms. This leads to abuses—sometimes in
the form of racial or sexual discrimination, other times in more
harmless ways. Of course a black person can nowadays apply for a
fellowship in Trinity College, but only after being "disinfected" by at
least one year of college life. Regarding the more farcical implications of university law, however, I've heard that once during an exam
a student caused consternation by invoking an obscure medieval
statute wherein it is written that examinees have the right to one
glass of ale. Confusion ensued, but in the end an enraged invigilator
did rush to a nearby pub to honor the statute in cause. The clerks,

---

*Needless to say the porters are the greatest snobs of all—a very English
phenomenon foreigners find hard to understand. This seems to be a pattern.
For instance, I have noticed that the people most obsessed with hierarchy in
British academia are the Ph.D. students.

however, took their revenge in due course, when after scouring the dusty pages of the book of statutes they finally decided to fine the student a large sum—for attending an exam without wearing a sword.

It was in this unusual environment that I studied relativity and cosmology, and wrote my first scientific papers. At the same time I was learning about the Big Bang riddles, I also found out that it hadn't taken long for scientists to come up with an answer: the inflationary universe. Soon after its proposal, Guth's theory was engulfed by a gigantic wave of enthusiasm within the scientific community, which has revitalized cosmology to this day. Inflation was created to solve the Big Bang riddles, and to some extent it does so remarkably well. However, inflation is not yet fact; it still awaits experimental confirmation. As I pointed out before, no one has ever seen an inflaton, the field that supposedly drives inflation. Until we do, there is a market for alternative ways to solve these riddles, and scope for much childish bickering among cosmologists.

And indeed, from my Cambridge vantage point, I found out very quickly that there was something about British physics that didn't like inflation theory. The British resistance to inflation, I would soon learn, was not entirely scientific. Scientifically, for sure, the Brits had their arguments. Inflation isn't really built on any piece of physics that one will ever be able to test in the laboratory. It lacks contact with "down to earth" physics. But I sensed there was something more. Maybe I had this feeling because I am Portuguese and therefore have something of an outsider's perspective. I began to suspect that the Brits didn't like inflation theory because their younger cousins across the pond had come up with it. And in the grand tradition of scientific competition, the British physicists weren't going to accept it until they were forced by incontrovertible evidence to do so.

Then again, they didn't have a theory of their own to offer in competition. Finding an alternative to inflation was no easy matter.

Regardless of what theory one developed, it either sounded very much like inflation or came disappointingly short of solving the Big Bang riddles. I began to feel that until you could offer a theory to compete with inflation, you had no right to criticize it. And the drive to come up with an alternative to inflation soon got me thinking deeply about these problems, month after month, year after year . . . but always to no avail.

Until one grey and damp winter morning, while I was walking across the playing fields of St. John's College. I was contemplating the horizon problem, and likely muttering to myself about how annoying it was. It may not have been very obvious to you how inflationary expansion can open up the horizons and homogenize the universe. Even less obvious, though, is why it is so difficult to solve the horizon problem without inflation. But to a trained cosmologist, the difficulty is there, and infuriatingly so. Inflation had won by default, simply because no other competitors had turned up in the ring.

Suddenly I stopped, and my mutterings grew louder. What if, in the early universe, light itself traveled faster than it does now? How many of those riddles would such a possibility solve? And at what price to our ideas about physics?

These thoughts fell from the sky, along with the rain, abruptly and without warning; but I immediately realized that such a possibility would solve the horizon problem. For the sake of argument, let's suppose there was a great revolution when the universe was one year old, and that before this, light was much faster than afterwards. Let's also ignore the subtle effects of expansion in the definition of horizons, which play a crucial role in inflation, but not in the standard Big Bang or VSL models. Then the horizon size at this time is the distance traveled by light—and that means fast-light—since the Big Bang: one fast-light year. If we did not know about fast-light, we would think that the horizon at this time was only 1 slow-light year across. And that is much smaller than the vast homogeneous region we can see nowadays: 15 billion slow-light years across. Thus the horizon problem. But if fast-light were very much faster than slow-light, it could be that 1 fast-light year is much larger than 15 billion slow-light years. Hence we would bring into contact at early stages all the vast regions we now observe to be so homogeneous. We could then open the doors to a physical process to explain the homogeneity of the universe. And we could do so without inflation.

I believe that this thought may have crossed the minds of many readers when I first described the horizon problem. It's so obvious. But I think one needs to be a professional physicist to recognize the tremendous heresy behind the suggestion and to be intimidated into rejecting it out of hand. Yet the idea is not as scandalous as it could have been. For instance, I did not entertain the possibility that someone could travel faster than light; nor did I propose that light could be accelerated. All I suggested was that the speed of light, still to be seen as a local speed limit, could vary, instead of being a universal constant. Believe me, I was being as conservative as possible, trying to stick to relativity's guns as much as I could, while trying to solve the horizon problem without inflation.

Of course, unlike inflation, this "varying speed of light" (VSL) theory still required serious modifications to the foundations of physics. It conflicted with the theory of relativity at line one. But I didn't regard this as a major drawback; on the contrary, I felt it might prove to be one of the model's most desirable features. I was immensely attracted to the possibility of using the Big Bang universe to gain insights into the nature of space and time, matter and energy, beyond our rather limited experience. Perhaps what the universe is trying to tell us is that at the most fundamental level physics is very different from what relativity teaches us—at least when we subject ourselves to the tremendous temperatures felt by the universe just after the Big Bang.

Having an idea, though, is but the barest beginning of any scientific theory. The flash of inspiration that filled me on that dank winter morning would have been utterly useless on its own. I knew that it required a mathematical theory to embody it and give it life. It was such an obvious solution to the horizon problem, and, as it turned out, to all the other riddles of the Big Bang. And yet it required revising the entire framework of physics, as laid down by Einstein at the start of the twentieth century. A tremendous enterprise loomed ahead.

THE BEGINNING OF THE JOURNEY was most inauspicious. Soon after my magical first encounter with VSL, I found that my ivory tower holiday was coming to an abrupt end. The high table threatened to tip me over into the garbage bin of unemployment, as my college fellowship was wrapping up and I was expected to find another job. Alan Guth's theory of inflation may have been motivated by his being under pressure to get a job, but VSL was actually inhibited by a similar constraint. I knew very well that if I suddenly started working full steam on such a deranged concept no one would ever employ me. VSL was such a risky and outrageous long shot that I would have found myself selling the *Big Issue* outside St. John's College.*

Besides, VSL was proving tricky to nail down. Every time I took it out of the drawer and tried to convert that beautiful insight into a concrete mathematical theory, disaster would strike. The formulae complained, shouted at me that they did not want a varying c, and spat out a stream of abusive apparent inconsistencies that always made me throw it all back into the drawer in humiliation and despair. I needed a collaborator; some things are just not made for lonely creation. I needed someone to bounce ideas off, to complement my failings, and to get me out of mental jams. Still, my attempts to get any-

---

*The *Big Issue* is a magazine written and sold by the homeless in Britain.

one to discuss VSL with me drew at best blank stares, at worse hysterical laughter and disparaging remarks.

To my eternal shame I have to admit that in the end I gave up and sailed through those tough days of professional uncertainty by suppressing VSL, choosing not to think or speak of it. It is not very romantic, but it is the reality. We all happen to be made of flesh and bones, and as a result suffer from material insecurities that often take over our lives. Perhaps the really shameful thing is the way society is structured, greedily geared towards conventional productivity. The surprising thing, sometimes, is that people still have any new ideas at all.

One afternoon in May 1996, I was walking along King's Parade, opening my mail as I went along, when deliverance arrived. I had been offered an advanced fellowship from the Royal Society. For me, this meant only one thing: Freedom! Do what you want, where you want, how you want, and rest assured that for at least ten years no one will bother you. I was beside myself with happiness; at last I could afford the luxury of being a scientific romantic again. It is such an expensive commodity these days.

At this stage, I already knew Andy Albrecht well; indeed, we had written three papers together. I decided to join him down in London—certainly seven years of the Cambridge psychiatric menagerie were more than enough for me. Strangely enough, I had never mentioned VSL to Andy; but then that summer something dramatic happened that would bind us together for years to come.

In the pretentious style typical of such institutions, Princeton University organized a cosmology conference to celebrate its 250th anniversary. Under the assumption that its illustrious reputation should be more than enough, the university supplied almost no money towards the conference's organization. As I walked past the pathetic enlarged replica of King's College Chapel (Cambridge) gracing the Princeton campus, I reflected on how often the United States copies just about the worst of British culture, including its academic arrogance.

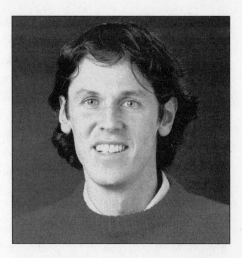

ANDY ALBRECHT, at about the time we first met.

But at least they had chosen the right organizer, Neil Turok, who promptly set himself the task of making the meeting as controversial as possible. I gather that Neil wanted to see blood on the floor: He organized the meeting in the form of a "dialogue" between opposing factions in whatever areas of cosmology still aroused passionate dispute. Although the term *dialogue* was really a euphemism for scientists trying to kill each other, this format worked very well, and to give you a taste let me describe what happened during one of the sessions.

One of the chosen topics was the evidence for the homogeneity of the universe as seen in galaxy surveys. In spite of what I told you when I described Hubble's findings, our best evidence for a homogeneous universe comes from the cosmic radiation. Galaxy catalogues still remain controversial. In fact, an Italian team has been analyzing these maps to claim that for all we know the universe is not homogeneous at all, but is a fractal. If this is true, burn this book, forget about Big Bang cosmology, and start crying convulsively.

Clearly this issue will be sorted out when maps containing larger populations of galaxies become available, which should be soon enough. But for the time being, "fractal people," as they have become known, play an essential role in cosmology: that of enforcing honesty. If you have crappy data and want to improve the way it looks, it is easy to start assuming homogeneity in the way you analyze the data, and it will look as pretty as if it had just emerged from plastic surgery. In this respect, fractal people have been instrumental in showing how circular some analytical methods in astronomy can be, assuming what they supposedly prove. I say this while sincerely hoping that fractal people are totally and irreversibly wrong.

At the Princeton conference, Luciano Pietronero, the Italian group's leader, defended his points brilliantly. The supporter of homogeneity unfortunately did not prepare well enough, assuming it would be a walkover, and he got a nasty surprise. To everyone's astonishment, Pietronero, while defending the unspeakable, actually managed to make much more logical sense.

Several other issues were discussed, with frequent similar upsets, and I recall fondly the debate over how fast the universe is expanding, that is, the current measurements of "Hubble's constant." Even though the contenders were close to reaching marginal consensus, this did not deter them from starting a hilarious contest of insults.

To my great surprise, given the stuffiness I had expected, the atmosphere was electrifying. The key issues were well identified and all sides were garrulously represented; Neil, meanwhile, kept the meeting from going overboard with the aid of a phenomenally large and noisy old-fashioned alarm clock that would go off with a deafening racket whenever a speaker went over time or tried to steal the limelight and hijack the proceedings.

It was against this backdrop that the issue was raised as to whether inflation really was the final answer to everything in cosmology. The day inflation was discussed was peculiar in that the arguments on stage spilled out into the audience, leading to a general free-for-all. Spurred on by the high hormonal level generated by this point, we all got into heated arguments, at times close to blows. As usual, the Atlantic seemed to be the dividing line in opinions.

At the end of this caustic day I got talking to Andy and another cosmologist, Ruth Durrer. Under the day's spell, Andy told us about his lifelong obsession, the need to find an alternative to inflation. As I told you earlier, one of the three seminal papers on inflation was also Andy's first scientific paper, written while he was a graduate student in collaboration with his adviser, Paul Steinhardt. Andy felt that his incipient scientific babblings could not possibly be the answer to all the problems of the universe. But if inflation was not the answer, then what was? Andy confessed that after all these years he felt at a loss. Everything he had tried either had not worked or had proved to be nothing but inflation in disguise, often a poorer realization of inflation. He asked us whether we had any ideas.

Ruth immediately tried to offer an explanation, but unfortunately she seemed to follow the Turok School of Solving Big Bang Riddles—she used the word "something" a lot, accompanied by a great profusion of gestures. I then briefly sketched the idea behind VSL. A stony silence fell: They thought I was joking, and that it wasn't even a good joke; indeed, it felt exactly like the sort of embarrassed silence that follows a particularly flat joke. It took them a while to realize I was serious. I was used to this type of reaction, so I didn't feel particularly upset about it. Only one thing seemed strange: I thought I could detect a faint gleam in Andy's eyes.

It is sometimes said that scientists spend all their time in conferences in exotic locations, wasting public money and having a good time. I wish this were true. More often than not, conferences are indeed a total waste of time and money, but they are also utterly dull. And yet every now and then a scientific conference does lead to new science, and this Princeton meeting was one such exception, in more ways than one. For the purpose of my story, it marked the turning point in VSL. I had finally managed to get a like-minded soul thinking about the problem.

Over July and August 1996 I lived in Berkeley and, as chance would have it, Andy was there too. However, Andy was busy writing his popular science book on the arrow of time, and I myself was working full-time on a different project, so I saw Andy only occasionally. But it was around this time that, somewhere overlooking San Francisco Bay, we agreed we would have a go at VSL together when we got back to London.

We were both approaching the project with some trepidation—we could already sense the total nightmare ahead. But the time felt ripe for it, or at least I was sufficiently immature to feel that way.

8 GOAN NIGHTS

I SPENT THE EVENING of December 31, 1997, at the Jazz Café in London's Camden Town, listening to my favorite jazz musician, Courtney Pine. His disheartened words, rendered in a deep voice as midnight struck, will forever be engraved in my memory: "Happy New Year to you all, as I am sure we're all happy to see the back of that last one. God, it was a hell of a bad year for me, but we've made it; it wasn't easy, but somehow we're still here, hoping the next one will be better—it certainly can't be any worse." I don't know how the rest of the crowd felt about it, but given what I'd gone through that year, I couldn't have agreed more.

The year had begun uneventfully enough. I had just moved to London the previous October, and I was still getting used to my new home, as well as to my new level of seniority. The latter had some definite advantages; for instance, I loved acting as an adviser to Ph.D. students. But some of the new responsibilities, in particular those related to administration, absolutely infuriated me. Why did people waste so much time with pieces of paper no one ever read?

In January 1997, I returned from a Christmas holiday in Portugal to find that Neil Turok had delegated to me the most depressing job on Earth—and possibly all other planets. I had been assigned the onerous task of directing the preparation of an enormous grant proposal involving some ten institutions across Europe. This meant pounds of forms to fill out and proposal text to write.

If you believe that cosmologists live in uninterrupted intellectual excitement, then lose your illusions at once. In reality, our financial survival depends on extremely bureaucratic institutions that manage scientific funding. These are controlled by ex-scientists well past their prime, so that these institutions wield a lot of power but otherwise function as a sort of intellectual scrap yard. As a result, instead of spending our time discovering new things, we have to waste long periods yawning at never-ending meetings, writing stupid reports and proposals, and filling in endless forms that do no more than justify the existence of these institutions and their senile staff. Indeed, I like to call grant proposal forms "old-fart certificates of existence," since as far as I can see their only purpose is to create a supposed necessity for these parasites. Why doesn't someone just set up a geriatric home for scientists who have stopped doing good science?

Deep in these intellectual doldrums I could do nothing but envy Neil, who had cleverly timed a trip to South Africa to provide himself with the perfect excuse to evade all that crap. Why had I not planned a trip to the South Pole for this time of year? Or to Andromeda? What a terrible lack of foresight.

No one believes me on this one, but I have a physical allergy to red tape. On these unhappy days I would go to Imperial College late in the morning, look with dismay at the dreaded forms lying on my desk, procrastinate until lunchtime, wandering along the just-after-New-Year empty corridors, and finally in mid-afternoon, in utter boredom, squeeze out of my brain a couple of uninspired phrases, trying to simulate an excitement I did not feel.

By the time I left Imperial, I would usually feel completely nauseated, full of self-loathing, ready to pick a fight in a pub. Isn't this a physical allergy? I wish I could convince a doctor to certify me incapable of performing red-tape tasks, whatever their description.

It was in this dark frame of mind that I would meet my girlfriend, Kim, at the end of the day for a drink somewhere in Notting Hill. By then I would feel so sick that I desperately wanted to clear my mind

KIM BASKERVILLE

of all that sewage by whatever means, and indeed after the second pint or so all those sordid forms would disappear down the anus of my brain. No wonder so many Brits are alcoholics.

I've always felt that there is happy and unhappy drinking. Like most Mediterraneans, I am naturally a happy drinker, and a glass of good wine is simply part of the general enjoyment of life. Northern European drinking is more often than not unhappy drinking—phenomenally voluminous, aimed at erasing from your brain a dreary day full of Protestant competence. Just the type of drinking I was in danger of falling into unless I did something drastic.

As it happens, Kim is also a physicist. We generally try not to talk about science, but on one of those evenings I felt sufficiently disgusted with myself to make an exception. In truth, I was just trying to shake off the viscous feeling of repugnance that any proper scientist feels when faced with bureaucracy. Unsurprisingly, I chose to go on a rant about the maddest thing I could think of, the varying speed of light theory, more to amuse myself than anything else.

I had told Kim about VSL before, but just in passing. Now I rambled on, trying to adorn my already lunatic idea with an even more psychotic coating. When she asked me why the hell the speed of light would change, I answered without hesitation that it was all a projection effect from extra dimensions. I spat this one out without thinking, but as it turned out, the idea made some vague sense.

One of the offshoots of Einstein's attempts to unify gravity with all the other forces of nature is the so-called Kaluza-Klein theories. According to these theories, we live in a world with more dimensions than the four (three spatial and one time) that we perceive. In the simplest Kaluza-Klein model, space-time is actually five dimensional: It has four spatial dimensions and one time dimension. If that is so, why don't we see the extra dimension? Klein had proposed that it's simply because the extra dimension is very small. If we ignore time for the moment, according to this picture we live on a three-dimensional sheet occupying a four-dimensional space. We are all flattened out so that we can never feel the larger space inside which we are embedded.

This may sound like an abstruse conception, and one may wonder why on earth the world would want to be like that. However, the first attempts at unifying all the forces of nature made use of this idea. Without going into detail, let me just tell you that the trick is to explain electricity as a gravitational effect along the fifth dimension. In the simplest Kaluza-Klein theories, gravity is the only force of nature—all else is an illusion created by gravity taking shortcuts through the extra dimensions.

Although Einstein himself was to devote a large part of his late life to this approach, most physicists never took it seriously, labeling it the apotheosis of deranged theorists. On this account, it may be useful to tell a story involving Kaluza, one of the creators of these theories. Kaluza made no apologies for being a theorist and felt irritated by the condescending tone with which nontheorists approached him and his views. We must recall that throughout the nineteenth century theoretical physics was a "poor man's" version of physics—the "real" physicist performed experiments. Indeed, the large number of Jewish scientists behind the great developments in theoretical physics in the early part of the twentieth century reflects nothing but this view of theory allied with ubiquitous anti-Semitism.

Against this background, Kaluza, a nonswimmer, set out to disprove the negative overtones of the term *theorist* by betting with a friend that he could learn how to swim purely by reading books. He collected a large number of books on swimming, and having satisfied himself with his "theoretical" grasp of the subject, he threw himself into deep water. To everyone's surprise, he could swim.

These days, Kaluza-Klein theories are no longer considered eccentric, and modern unification theories make use of them as a matter of course. What came into my mind on that evening, while talking to Kim, was the possibility of using them to implement VSL. The idea was certainly amusing enough to make me forget those disgusting forms.

My argument relied on the fact that in some Kaluza-Klein theories the fourth (extra) spatial dimension is not only small, but is curled up. In this particular picture we live not on the surface of a thin sheet (as explained above), but on a wire. The "length" of this wire represents the three extended dimensions we experience; its cross section is a very small circumference, and it represents the extra spatial dimension that we cannot perceive. This may be difficult to visualize, so take a look at Figure 8.1. It was not my idea, and most modern Kaluza-Klein theories use curled-up extra dimensions.

But now suppose that light rays move along helices, rotating around the circular extra dimension as well as moving forward along the wire, that is, along the three dimensions we can see (Figure 8.2). This unusual geometry of the universe implies that the fundamental and constant speed of light is its speed along the helix, not the speed we actually observe, which is its projection along the three-dimen-

Curled up
extra dimensions

Extended, visible, three spatial dimensions

FIGURE 8.1   The Kaluza-Klein wire world. According to this conception, the world is a higher dimensional wire, with its "length" made up of the three spatial dimensions we can observe, and the extra dimensions curled up.

FIGURE 8.2   Light propagation in the Kaluza-Klein wire world. If light snakes around the wire in a spiral, its actual speed is much larger than the three-dimensional speed we observe. If we could force light to move straight along the length of the wire we would observe a larger speed of light.

sional "length" of the wire. The angle, or turn of the helix, relates the two speeds. If this angle could vary, according to some prescribed dynamics, we would observe a varying speed of light, as a projection effect, within the framework of a theory in which the fundamental, higher dimensional speed of light was in fact a constant.

The trouble was explaining why the observed speed of light appeared to be constant, which in this setting amounted to fixing the angle of motion creating the helix. My idea was that this angle would be quantized, as are the energy levels of an atom. Quantum theory tells you that most quantities can exist only as multiples of indivisible basic amounts, the quanta. Thus the energy of light of a certain color must be a multiple of a given small amount of energy, corresponding to the energy of a single photon with that color. Similarly, the energy

levels of an atom are organized like the steps of a ladder, with electrons taking up orbits that have to fall within a given spectrum of
possible choices.

In the same vein I was hoping that the angle of helical light in this
Kaluza-Klein theory could adopt only a discontinuous spectrum of
values. Each allowed angle would lead to a different speed of light as
perceived by us, but it would take a lot of energy to cause a jump
between one quantum level and any other. Hence only when the very
high energies of the early universe were achieved could we unwrap
the helix and let the universe experience a larger speed of light. Or
such was my wishful thinking.

I did not know it at the time but this idea was not entirely new. It
had long been realized that in Kaluza-Klein theories the constants of
nature (such as the electron charge or Newton's gravitational constant) as seen by the larger space are different from those perceived
by the three-dimensional beings that we are. In general the two sets
of constants are related by the possibly variable size of the small
extra dimensions. The problem with this approach is that it runs into
the impenetrable wall of unknown quantum gravity physics, so that
predicting the behavior of such a model is more guesswork than hard
science.

My model was slightly better (the idea of quantizing light's helical
angle is not bad), but it ran into other problems. For instance, the
lowest energy state should actually be the one in which there is no
rotation in the extra dimension, that is, straight nonhelical motion.
But this would correspond to the largest possible three-dimensional
speed of light, and what I wanted was precisely the opposite: for the
speed of light to be larger when the universe was hot, not when it was
cold. There were ways around this difficulty, but they were very far-
fetched.

I never fully pursued this idea, but it got me thinking seriously
about VSL at long last. It showed me that there were ways, albeit
imperfect, of implementing the theory within the framework of

known physics. Not only had I gained a collaborator the previous summer, but just as importantly I was also gaining confidence.

A few days later, Neil returned from his trip to find that I had done pitifully little on the grant proposal front, and worse, that what I had done was next to useless. Kim worked with Neil at the time, and the next day she returned from Cambridge reporting with great amusement that Neil was extremely unimpressed with my performance, having delivered himself of the opinion: "You can't trust João with any administration."

IN THE WAKE OF this early idea, during January 1997, I devoted all my time to VSL. To some extent, VSL became a way of disinfecting myself from that polluting start of the year. Finally I had the security, motivation, and confidence to work on VSL full steam.

But in these pursuits I was still mostly on my own, despite what Andy and I had agreed upon the previous summer. Andy was very excited about VSL, and listened gleefully to all the nonsense I threw at him, but at the time he was far too busy to do any science, of whatever description. He was becoming a martyr to the red-tape pressures I have just described, and it got to the point where he had to lock himself into his office to get any science done. Even then he would immediately be derailed by secretarial requests as soon as he had to go to the bathroom. I suggested that he keep a chamber pot in his office, but I don't think he ever took me up on this precious piece of advice.

Naturally, I felt a bit impatient with him; after all, I had come to Imperial to do science, not to be buried in a pile of bureaucratic toilet tissue. But I know that he must have felt even more frustrated than I did. In addition, his personal life was very complicated, and growing more so.

Andy had moved to London from Chicago, with his wife and three children, to take up a tenured position at Imperial College. He soon found that British scientists are expected to live like monks: in extreme poverty, preferably without a family, and feeling as miserable as possible. Underlying this conception is the English taboo of discussing financial matters—money is not an issue a scholar should raise. Presumably this goes back to the days when all British scholars were wealthy gentlemen. As the social composition of academia widened, the new working and middle-class arrivals copied all the worst aspects of the upper class, in compliance with British custom and tradition. Every time I mentioned low academic wages during meetings, people started shifting their buttocks about on their chairs in clear physical discomfort. How vulgar of me to mention money—only some Latin wog could do something so crass.

The English attitude can be summed up in the maxim that the solution to the world's famine is for everyone to starve, not for everyone to eat. Not only do people seem to like being miserable, they also hate anyone who appears to be successful and happy. I remember Cambridge University giving well-funded students from the continent a hard time, arguing in writing that if English Ph.D. students live in poverty, why shouldn't the others as well? When I bought a new flat, a relative of one of my students, who had previously always been friendly toward me, suddenly became openly hostile. He later admitted that he couldn't stand the fact that I no longer lived in the same sordid conditions in which he still did. Britain is the only country in the world where the vast majority of uneducated people *want* their children to be uneducated too: "If it was good enough for me, then it's good enough for him."*

To people like me, who have no dependents, the issue of low academic wages is actually not terribly pressing. But if you have a family, and even worse, live in London, then joining academia entails a seriously low standard of living. Coming from the United States, the

---

* When I told the many stories I know illustrating this point to a South African social worker, she thought I was lying. The people in the slums of Johannesburg with whom she worked might be alcoholics and even criminals, but they did all they could to educate their children and get them out of poverty's vicious circle.

Albrecht family never recovered from the shock—after all, hanging on in quiet desperation may be the English way, but Americans are simply not prepared to put up with all that metaphysical crap. I know for a fact that in all the time Andy spent at Imperial, he never stopped applying for jobs back in the United States, hoping to take his family away from that nightmare. In addition to family pressure he also had the administrative chaos of Imperial hanging over his head. If I felt impatient, I can only wonder how he felt.

Nonetheless, whenever time allowed it, Andy listened to my increasingly persistent VSL ramblings with interest, even with envy, although he seldom took an active role. But then one day, sometime in February 1997, Andy called me to his office and locked the door. He rather dramatically announced that the time had arrived for us to work on VSL properly, and sod all the rest.

I had seen these outbursts before in other scientists, and had even had them myself. You suddenly feel that the only reason you endure such low pay is that you love the work you do. But then you realize that all your time is being devoured by paperwork and science administration. So you reach a point where you explode and tell yourself that if you are going to give priority to all this crap you may as well work in a bank and be paid properly. Piles of forms usually find themselves being flushed down the toilet in the wake of these epiphanies. And then you sit back, relaxed and happy, fully reconciled with the universe, and start doing research intensely, ignoring all the messages left on your phone by those idiots up in the Sherfield Building. A crimson wave of freedom ripples through the universe auguring the advent of the Golden Age . . . until reality ferociously catches up with you.*

You don't do science by decree, but indeed a very prolific nine-month period followed this "historical" event. I started to go regularly

---

*The Sherfield Building is Imperial College's administration venue. It absorbs large sums of money and generates tons of useless paperwork. I have suggested that it would be a major improvement if they were still allowed to

to Andy's office and we would brainstorm until I had a headache. A lot of what we argued about was nonsense, but of the sort that eventually led to promising new avenues. My early Kaluza-Klein idea was soon abandoned in favor of what we hoped were better-defined and less impetuous approaches. Slowly, we began to drift towards something that vaguely resembled an actual theory. But was it the right theory?

At the end of each of these brainstorming sessions, Andy always wiped out everything we'd scribbled on the blackboard. VSL became classified, as Andy worried that someone might steal our idea. Apparently, he'd had some bad experiences in this respect early in his career, and now took all possible precautions. I've never been that paranoid, but this was a welcome change nonetheless. A few months before, my idea had been too inane to be graced with a comment. Now all of a sudden it was so precious it had to be locked in a safe until the project was mature enough for publication—when its paternity could be firmly established. As a result, during this crucial period of its development, VSL was pretty much confined to Andy and myself.

But there was also something else that was refreshingly different, and that was Andy's attitude to the "unknown." A few months before, I had been permanently getting stuck: Whenever I put a varying c into the usual formulae of physics, the whole thing just crumbled into mathematical nonsense. Confused and disappointed, I would just give up. Somehow, having someone else to discuss things with was just what I needed to help me understand that such mathematical disasters were not absolute indications of genuine

---

waste all that money, but were prevented from doing any "work." More radically, and drawing on early leanings, I have at times considered launching a devastating terrorist strike against staff and building. IC's average IQ would increase significantly and a higher quality of both teaching and research would inevitably result.

inconsistencies but merely reflected the limitations of the available language of physics. With this in mind, it was much easier to recognize what the collapsing formulae were trying to tell us. And to build new formulae that could accommodate a varying speed of light consistently.

Crucial to these breakthroughs was Andy's reckless approach. His attitude was, the heck with all of it, let's just come up with something that has interesting cosmological implications. If those string theorists are really as clever as they think they are, they'll work out the details for us later.

OUR DISCUSSIONS FOCUSED on the cosmological implications of a varying c: We wanted to produce a new model of the universe that was capable of explaining the Big Bang riddles, yet at the same time was radically different from inflation. However, it was clearly not enough just to assert that the speed of light in the early universe was greater than it is now, and that that solved the horizon problem. Logically, a varying speed of light must have many more implications for the basic laws of physics, and ultimately for cosmology. We needed to find a mathematically and logically consistent way of implementing VSL. In other words, we needed a *theory*. And what about the other riddles of Big Bang cosmology? After all, to some extent the horizon problem is just a warmup for more serious problems.

And so we started wondering more generally what else would change if c varied. This is a very wide-ranging question, and it led to a long process, unfolding throughout the following months, in which we gradually scanned the effects of a varying c on most of physics. We discovered that a changing c would have deep consequences for every single law of nature.

New terms would necessarily appear in most equations, terms that we code-named the "c-dot-over-c" terms. This expression became a running joke between Andy and me, and merely refers to the mathematical formula for the rate of change of the speed of light.* Corrections to the usual formulae of physics had to be related to this rate—the dreaded c-dot-over-c terms. They became the focal point of our research. What were these crucial terms, and what new effects would they predict?

Soon I was so totally immersed in the nightmare of working out the "c-dot-over-c" terms that I no longer knew what to do with them. We were making progress, but we were also in a terrible tangle. There were so many different avenues: How could we tell which one was the most promising? It became an embarrassment of riches and as such a real nightmare. There is not much point in my telling you about these early approaches—suffice to say that there were *many* of them, and that the vast majority were truly crappy dead ends. While red tape piled up on our desks, every now and then mysteriously falling into the garbage bin, we puzzled over VSL, more often than not feeling a bit at a loss.

Finally, in April, the need to refresh my mind became so over-whelming that I decided to forget about it all for a while and disappear from London with Kim. We chose to escape to Goa, a beautiful part of tropical India that I had always wanted to visit. Goa was formerly a Portuguese colony, but its all-powerful colonial lords were driven out in the early 1960s by the Indian army. Various speed records were beaten during the retreat, in one of the few episodes of Portuguese colonialism I find amusing. Nonetheless, the Portuguese did leave behind a modicum of improvement, for example a half-decent edu-cational system, in sharp contrast with that in most (although not all) of the rest of India. Indeed to this day you get the feeling that Goans wanted to be independent rather than to become part of India, and they preserve a distinct cultural identity containing numerous Portuguese elements. Some people still speak Portuguese, or even more embarrassingly sing the *Fado,* a Portuguese version of the blues.

---

*More precisely, the expression $\dot{c}/c$.

No sooner had the Portuguese left than the California hippies arrived, and ever since Goa has had to endure generation after generation of Western fringe lunacy. Semipermanent colonies are now established, and Goa is firmly on the map of nomadic wanderings for all self-respecting peace-and-love followers. In 1997, when I first visited Goa, the rave culture was at its height, with all-night beach parties striking at full moon, and Goan trance music blasting out into the Indian Ocean and the rest of the universe. Such was the place I went to relax my brain cells.

Predictably, Anjuna, where we stayed, was quite a zoo, both in the literal and metaphorical senses. In the former there was an abundance of stray cats, semirabid dogs, cows wandering on the beach, monkeys playfully sitting in bars, goats, pigs, etc. We soon acquired faithful dogs, Goan dogs being desperate to find an owner, mainly to protect themselves from other dogs. As for the metaphorical zoo . . .

While we lounged about on rugs in an Afghan "restaurant," ravers launched impressive displays of fireworks in celebration of the birthday of the ancient Afghan granny of the house. Rave music was interrupted while the granny's special wish was granted, and Pink Floyd was played on the sound system. Over breakfast, we were serenaded by great tirades on ethics and other branches of philosophy by a totally deranged French girl, whom we quickly dubbed Simone de Beauvoir.

Rave parties on the beach extended until sunrise, punctuated by the odd police helicopter passing over, pointing floodlights at us in clear warning that insufficient bribes had been paid. In return, the crowd would aim their lasers at the helicopter, projecting little red hearts all over it.

Hippy remnants played the flute to irate dogs, who rolled around barking and biting at each other in the middle of bars and restaurants. Waving goodbye to the setting sun on the beach started to look like the most normal thing in the world.

Funnily enough, in sharp contrast to the naked hippies living in treetops and the ecstasy-fueled ravers, among the Goans themselves one could sense an undertow of Portuguese *brandos costumes,* or mild manners, a fossilized old-fashioned way of life that has not survived to this day in Portugal itself. I made friends with some of these people, such as Sr. Eustaquio, the proud owner of a large Portuguese speaking parrot, and Francisco, owner of the restaurant Casa Portuguesa and an expert *Fado* singer. I remember fondly the exquisite pleasure of returning home from his restaurant, at five in the morning after a tropical storm, singing the *Fado* at the top of my lungs, thousands of miles away from Bairro Alto (the bohemian district of Lisbon), waking up all the Goan fauna.*

Although using one's brain appeared to be discouraged in this peculiar environment, I must say that mine worked better than ever. As I relaxed, a few VSL breakthroughs suddenly came to me. Naturally I only jotted them down briefly, waiting until I was back in England to work out the details—Goan nights are not exactly conducive to performing taxing calculations. But slowly, in the background, a small pile of interesting ideas started to materialize. I recall sending Andy a postcard of a palm-fringed beach, reporting that I was spending all my time working on those "c-dot-over-c" terms. He must have thought it was a joke, but it was at least partly true.

Late at night, using God's toilet—the only one available in most Goan bars—I would accidentally look up, through the palms, into the heavens. With little or no electric light to corrupt it, the darkness of the Goan skies only left room for the infinity of stars. I know that observing the universe while pissing may not be the most poetic setting, but the shock was always the strongest, as the full weight of the universe fell into my eyes. From a faraway sound system, I could hear

---

*Kim was no more impressed with my *Fado* performances than the Goan animals, and kept up a barrage of disparaging remarks until I eventually resorted to singing the *Fado* while swimming far out at sea, where nobody could hear me.

the rave cliché broadcast in an electronic voice: "When you dream there are no rules, anything can happen, people can fly."

WHEN I GOT BACK to London, suntanned and happy, VSL had entered a new phase. The short notes I had scribbled in Goa had paid off, and soon what had begun as a fun insight was blossoming into a proper mathematical theory, no matter how crazy. Little by little, my classified meetings with Andy led to more concrete paths through the physics landscape. The "c-dot-over-c" terms emerged from the labyrinth, and the new physical effects we sought began to crystallize.

What else would change if c varied? Some of the consequences were quite dramatic indeed. Perhaps the most alarming discovery was that the conservation of energy—a central dogma of science since the eighteenth century—was violated. A varying speed of light allowed for matter to be created and destroyed.

This may sound strange at first, but it is easy to understand. Early in the twentieth century, scientists realized that the conservation of energy was simply another way of saying that the laws of physics must remain the same at all times. This more abstract approach should actually be taught in schools, for otherwise the conservation of energy will always appear to be a miracle. In fact it reflects nothing but the uniformity of time—we change, the world changes, but the laws of physics remain the same forever. A few mathematical calculations, and conservation of energy trivially follows.

By changing the speed of light, we were violating this principle by forcing the laws of physics to change too. Indeed, the speed of light is ingrained in the actual formulation of all laws of physics, at least since the advent of the special theory of relativity. Therefore, it was not really surprising that the conservation of energy went out the window. We were allowing the laws of physics to evolve over time, in perfect contradiction to the fundamental principle underlying the conservation of energy. In VSL it is only logical that energy should *not* be conserved.

That much had already become obvious to me, by other means, in one of my Goan scribblings. In fact, I cannot believe I hadn't noticed it earlier; anyone with a basic knowledge of differential geometry would have seen it at once. Einstein's equations tell you that matter curves space-time, and that its curvature is proportional to the energy density. But curvature must satisfy a set of identities called the Bianchi identities, which are a mathematical necessity and have nothing to do with general relativity. They constitute a statement akin to $1+1=2$, and are valid for any space-time whatever its curvature. But if curvature is proportional to the energy density according to Einstein's field equation, then what do the Bianchi identities imply about energy? It turns out to be nothing but energy conservation.

But wait: I told you that curvature is proportional to the energy density—that means that curvature equals the energy density multiplied by some number. What is this number, the so-called proportionality constant? Hidden inside this constant lurks the speed of light. If the proportionality constant is indeed a constant, then Bianchi identities imply energy conservation; but if it isn't, as would be the situation if the speed of light were not constant, then these identities actually require *violations* of energy conservation. The full argument is somewhat more complicated than this, but what I have just told you nevertheless gives the flavor of one of my Goan notes. I had found out VSL implied that energy could not be conserved.

We therefore had two lines of reasoning pointing to the fact that energy would not be conserved under VSL. And when we performed the necessary calculations to find out how much violation should occur, it all matched—the two approaches agreed. But then Andy and I made an incredible discovery.

Our equations revealed that the way in which the total energy content of the universe would change was determined by the curvature of space. If gravity were curving space in upon itself to create a closed universe, energy would evaporate from the universe; if space were becoming saddle-shaped, creating an open universe, then energy would be created out of the vacuum. Now, according to Einstein's famous equation $E = mc^2$, there is an equivalence between mass and energy. Hence mass would be removed from a closed universe and it would be created in an open one.

This had a dramatic consequence. Recall that a closed universe is one with a mass density above the critical density that characterizes a flat world. As a closed universe lost energy, its mass density surplus would be shaved off, and the universe would be pushed towards a flat, or critical, configuration. An open universe would gain energy, and that means its mass density would also increase. But we have seen that in such a universe the density at any given time is below the critical density. Therefore, due to violations of energy conservation, any deficiency in its mass density with respect to the critical value would be restored, so that the universe would again be pushed towards the critical, flat case.

Under our scenario, then, a flat universe, far from being improbable, was inevitable. If the cosmic density differed from the critical density characterizing a flat universe, then violations of energy conservation would do whatever was necessary to push it back towards the critical value. Flatness, rather than being a tightrope, became instead a gorge into which all other possible universes would happily fall. And in a flat universe, matter was neither created nor destroyed. We had discovered a new, noninflationary valley for flatness.

From this point onwards, Andy and I were on a high. In setting out to solve one cosmological riddle, the horizon problem, we had stumbled upon a solution to another, seemingly unrelated one: the flatness problem. Gradually, in the middle of these physics storms in Andy's office, it dawned on us that we were getting more than we had bargained for. The deeper we got into the physics, the more cosmological problems we seemed able to solve, some rather unexpectedly.

Most obviously, we could explain the origin of matter. Out of seemingly abstruse properties of the theory, such as the possibility that matter is created due to a varying c, we found—much to our surprise—an explanation for where all the matter of the universe had come from. This isn't one of the traditional Big Bang riddles, but for me it is a far more important question, something everyone must have asked at least once: How did the universe come into being? VSL provided the very answer.

THESE EARLY SUCCESSES sparked a period of very hard work through May and June 1997. We knew we were finally on the right track, and that pushed us further and further. In those days, I was so excited about VSL that I often stayed in my little office at Imperial College until very late, sometimes four or five in the morning. I worked hard on the emerging details of the new theory, uncovering more exciting features at every step. During this period, I became friendly with some of the Imperial College security guys, who no doubt must have thought me an oddball. There was also a student who worked all night who truly looked like Count Dracula. The first time I saw him, wandering at the other end of the corridor well past two in the morning, I thought that perhaps all the excitement was having a detrimental effect on my mental health.

Unfortunately, these adventurous flights do not happen often in science. But when they do, they're unique, a massive adrenaline rush difficult to match in any other way. I have wondered whether this might be the reason scientists are often such nerds. Perhaps after these extreme intellectual experiences, the more proverbial pleasures—eating, drinking, chatting to friends—feel dull by comparison. This may well be the reason so many of us become the ultimate social suicides.

Certainly I was on the verge of becoming a lonely nocturnal beast as I made my way home late at night, walking through empty streets in an eerie silence seldom felt in such a big city. It is not widely known, but some parts of central London are home to a large population of foxes, who take over the town after hours. I myself did not know this until I ventured out into these unearthly nights. But as I went home exhausted and with my brain in a muddle, I would suddenly find myself in the company of these creatures, parading their distinctive bushy tails, unhurriedly going about their lives. Occasionally, one would stop to stare at me, presumably wondering what sort of nocturnal beast I was. It would then slip into some garden, only to reappear several streets away, using shortcuts known only to foxes, in a parallel city beyond our sensible world.

Some of what I worked on during these foxy nights was just the grotty, dirty details. For instance, we had to know by how much the speed of light had to change, and how it would change. In these early days, Andy and I envisaged VSL as a cosmic cataclysm in the early life of the universe, close to the Planck epoch I mentioned before. As the universe expanded, it would cool to a certain critical temperature, at which point the speed of light would suddenly change from a very high value to a very low one. We pictured something like a phase transition, a bit like water turning into ice as the temperature drops below freezing point. Likewise, the cooling, expanding universe would cross a "freezing" temperature, above which light would be much faster and "liquid," and below which it would crystallize into the "slow" icy light we can observe today. Later on, we found that this was only one among many possibilities, indeed the simplest one, but let's stick to it for the time being.

The challenge was then to place conditions upon this phase transition so that we could solve the horizon problem. The answer was that for a VSL phase transition occurring at the Planck time, the speed of light would have to drop by more than a factor of 1 followed by 32 zeros if it was to causally connect the whole observable universe. If you thought that 300,000 km/s was fast, add on 32 zeros and you get a truly incredible speed. In fact this was the minimum requirement, and we were so puzzled by this large number that we decided to favor scenarios in which the speed of light in the Planck epoch is actually infinite. Under these circumstances, the whole observable universe was once well in causal contact, established by zippy light.

Also in this scenario, as soon as the universe came out of the phase transition, it found itself walking the tightrope of flatness again. But this happened after flatness became a gorge, while the speed of light was dropping. The question was now to find by how much the speed of light had to change so that this primordial balancing pole gave the universe enough security to endure the tightrope of flatness in its ulterior life. The answer was precisely the same as the one we had got before, in order to obtain a solution to the horizon problem. The primordial speed of light would have to have more than 32 zeros added to its current value. At the time we did not know it, but this was far from a coincidence.

And so forth . . . as these long nights spread themselves in front of me, I found a wealth of details materializing at long last. But we had discovered two fundamental and essential things at this point—that VSL led to violations of energy conservation, and that this solved the flatness problem in addition to the horizon problem. There were also a couple of bonuses, for example an explanation for the origin of all the matter in the universe. But one crucial element was still missing: What about the cosmological constant?

It was obvious from the start that there had to be an interesting interaction between the cosmological constant and a varying speed of light. After all, if the speed of light is demoted to a wild and variable beast, why should the vacuum energy remain a rigid constant? And indeed it soon became obvious that if c were not a constant, the energy stored in the vacuum could not remain immutable either. The vacuum energy can in fact be written according to that strange geometrical object introduced by Einstein, Lambda, but then on closer inspection one finds that the speed of light also features in the formula. In general, one finds that the vacuum energy increases if the speed of light increases too.*

Conversely, if the speed of light decreased in the early universe, the vacuum energy would sharply decrease. This energy would then be discharged—into all the matter and radiation in the universe. And so VSL could do what the cosmological expansion, even inflationary expansion, could not: get rid of that domineering vacuum energy. Recall that the problem with the cosmological constant is that the vacuum energy is not diluted by expansion, in contrast with what happens for matter and radiation. For this reason, the vacuum energy should dominate the universe, and quickly so, unless we can find a way to suppress it brutally in the early universe. VSL supplied precisely such a mechanism—a way of converting any vacuum energy into the matter of the universe, thus leaving the universe to expand into old age without the threat of sterile domination by the nothing. We had just found a way to exorcise the cosmological constant.

Naturally things were not quite as simple as I have just painted them. We knew that our mechanism was not perfect, and that it solved only one aspect of the cosmological constant problem as particle physicists had been recharacterizing it over the last few decades. But until this point, there had been times when I could not help feeling that to some extent VSL was a pedantic exercise. We were solving problems that had already been solved—by inflation. We had had a few beautiful surprises, but in essence what else was new, apart from the idea of VSL itself? All of a sudden, the whole VSL landscape had changed. We found that VSL could beat back the threat of the Lambda beast. Inflation could not solve the cosmological constant problem, but VSL could.

By the end of June 1997, we were ready to hit the world with our cherished monster. A lot of work had been done, and a phenomenal

---

*More precisely, it appears that the vacuum energy is proportional to Lambda times the speed of light to the fourth power.

pile of notes had accumulated. I was more excited than ever, and Andy also seemed very pleased with our VSL theory.

But then all of a sudden, Andy got cold feet. Gradually, and for no apparent reason, I sensed that he had started to feel uneasy about our mad project. What I didn't realize at the time was that Andy's trepidation would nearly derail VSL altogether.

9   MIDDLE AGE CRISIS

IN HINDSIGHT I CAN SEE that VSL was born from a huge manic-depressive swing. Until June, Andy and I had been on an emotional high, bathing in never-ending energy. But we are all creatures allergic to eternal happiness, so this soon had to come to an end. A darker season lay ahead.

As the end of June approached, we had enough material for not just one paper, but several. Truth be said, this was partly because we had found not one but two versions of VSL, one more complex but founded on firmer ground than the other. The physical content of both theories, however, was very similar. We decided to write a first tentative article to mark territory, a bit like dogs pissing on a wall. Unsurprisingly, we chose the simpler but vaguer version of the theory.

In July, Andy and I would be going to the Physics Center in Aspen, Colorado, for a two-week stay. The center is organized along very unusual lines: The number of talks or presentations in any given program is reduced to a minimum and emphasis is put on informal interactions between scientists. In practice, as Andy warned me, there is always the danger that your ideas could be stolen. People who work very hard, but are otherwise devoid of talent or imagination, hang around such places tapping "informal" conversations, and making successful careers out of it. Every year a renowned American university awards a prize to the best paper based on someone else's idea.

Andy felt that this would be a perfect occasion to discuss VSL with a wide community, but he was adamant that we needed to insure ourselves by first writing a paper and placing it on a Web archive, such as http://www.astro-ph.soton.ac.uk. This would establish our informal paternity, and we could then use Aspen to discipline our bastard son.

For the record, Andy wrote the abstract, the blurb, and the conclusions, and I selected stuff out of my notes to make up the central sections. This may sound trivial, but it took ages to do! Andy suddenly started to become very morose, a fact that I attributed initially to his red-tape pressures. But gradually, I realized that there was more to it than that.

A few days before I was due to leave for Aspen, Andy stayed late at Imperial so that we could at long last finish our paper. Over dinner at a nearby restaurant, Andy came out with it. He admitted that he was afraid of submitting the paper. He wanted to sit on it for a while.

I had seen this phenomenon before: A few days before the submission of any scientific article there is always one author who gets the shits and starts making excuses to delay it. It's a common psychological effect, a counterpart to stage fright. But our situation was different—coming out with VSL *was* scary—and we probably *should* have been afraid. After all, we were doing nothing less than demolishing *the* pillar of twentieth-century physics: the constancy of the speed of light.

Perhaps for this reason I got frightened by sympathy and made a decision I would later regret: I agreed to wait. But this meant delaying submission until after Aspen, which made me uncomfortable. I was starting to feel that VSL was suffering from a lack of external feedback. The project had now been gestating for over six months, fully shielded from the outside world. This could not be healthy; you normally consult your colleagues at every stage of an idea's development.

The only exception had been a conversation I had had with the head of our group, Tom Kibble, widely known for his dry, cutting views. I went to his office and told Tom that Andy and I were looking for an alternative to inflation. He promptly replied, "It's about time." I smiled and started explaining the set-up of the horizon problem. He said, "That's quite reasonable." I then described the VSL solution to the horizon problem. He said, "That's less reasonable." When I moved on to explain the intricacies of energy conservation violations, Tom fell asleep. I left his office while he snored away, happily drifting on a different horizon.

I told Andy about my fears that we would be missing important input by not discussing VSL with other cosmologists in Aspen, but Andy said there was nothing we could do.

IF THE PRINCETON CONFERENCE the previous summer had been intellectually galvanizing, my stay at Aspen was the ultimate yawn. Indeed, the VSL depressive swing started there. Although the place is supposed be a haven for the informal exchange of ideas, it's exactly the opposite. Perhaps because of the highly competitive nature of U.S. science, Aspen is a place where you poignantly notice people interrupting their scientific discussions and changing the subject when you join their "informal" chat in the gardens. On a couple of occasions, I overheard some of what was being said, and sure enough, papers on those matters appeared soon afterwards, with suitably big splashes. When Andy arrived and we started talking about VSL, I saw him in turn change the subject when other people joined us. And this was the cream of U.S. cosmology in action.

I found the atmosphere quite uncongenial; this was a different world from the free-for-all I had grown used to in Britain ever since my Cambridge days. I got on very well with everyone at Aspen, so I did not feel that people were excluding me for personal reasons—

they were just doing what they felt they had to. But when it came to my turn, and I saw Andy hiding our stuff, I felt sickened by it.

Surely, unpleasant as this atmosphere might be, it pays off. Objectively it reflects nothing but the higher intensity of U.S. cosmology, allied with more unforgiving competition. At any given time, most active cosmologists will work on the same inflation-based problems, whatever is considered to be the flavor of the season, and not surprisingly they find themselves in a crowded and cutthroat environment. The collective advantage is that when the hot topic just happens to be a matter of fundamental importance (and you can never be sure of that), you have the full weight of an enormous community working on it—so, statistically, the system has to work. It leads to such a phenomenal output that it's bound to contain some work of real quality. But at the same time it's difficult to identify any sense of fun or freedom.

This was the first time I had experienced in such depth this method of doing science, and it came as something of a surprise. After all, the image U.S. science likes to broadcast is that of the freewheeling individual. Richard Feynman once wrote a beautiful piece aimed at those wanting to become scientists. He expressed his regrets over there being less and less room for innovation in science, but strongly urged us to rock the boat. He said we should all follow our noses, try our own thing no matter how crazy, and feel the loneliness of originality, even if that meant a brief career. He warned us that we should be ready to fail, and that we would indeed mostly fail if we imposed our individuality on our science. But he still felt the risk was worth it.

Feynman himself was a good example of what he advocated—a fuck-everyone-else-I'll-do-my-own-thing-and-don't-care-about-what-you-think kind of scientist. He has become an icon of U.S. science, and yet its prosaic reality is entirely different: It's a world in which young people are encouraged to work en masse on the same mainstream problems, without the balls to move away from the madding crowd. As with scientific red tape, my feeling is that if you're going to do science in this way, you might as well go and work in a bank.

My experience in Aspen was particularly disappointing because most of the other times I had visited the U.S. I had truly loved it. I had always found people interactive, open, and intense, in total contradiction with what I saw in Aspen. Perhaps these visits were to microclimates—secluded environments. Or else Aspen itself is the exception. How can these two sides of the coin be reconciled?

Maybe the answer is that in science, as in everything else, the United States defies generalization. It contains the best and the worst of all possible worlds, both in large amounts. I spent five months with Neil's group at Princeton, and made several visits later, always to find one of the most stimulating environments I have come across. I also spent two months in Berkeley, only to find a group of semi-deranged people, constantly sniping at each other, keen to suppress new ideas.

From this wide-lens perspective, what I found in Aspen is both typical and also an affectation peculiar to a minority. To make a comment about U.S. science is like trying to make a general statement about music. Some music you like, some you don't. . . . Do we have to like all music in general?

Regrettably, people are often most proud of their most appalling attributes, and indeed many American scientists appear to be more appreciative of bandwagons than of their Feynman legacy. Of course, they are not alone in this. I once met a girl in New York who was thrilled to find out I was a physicist; but she became terminally disappointed upon hearing that I lived in England and harbored no ambitions to move to the United States. She simply couldn't understand that. When I asked why, she tried to reply with an example, but she couldn't remember the name of the physicist in point. She asked me, "Who was the physicist who was better than Einstein, but never came to the U.S. so he never made it?"

To this day I have no idea who this mythical character might be. But her views on Einstein and on American virtues are more than laughable. Poor Albert: as if he had derived his greatness from having moved to the United States! At the time he crossed the Atlantic his best work was finished anyway, and he had already received Nobel Prize–level recognition. He moved only because of the Nazi regime, which he had antagonized from the very start, at a time when everyone else, including many rich Jews, was still trying to strike a compromise. Indeed, his political outbursts quite often caused embarrassment, and in this respect Einstein sometimes reminds me of Muhammad Ali. Unsurprisingly, Einstein was unceremoniously expelled from Germany in 1933, with all his worldly belongings confiscated and amid rumors of planned attempts on his life.

Einstein was received in the United States with open arms at a time when he desperately needed such hospitality.* Perhaps if that girl had seen it in this light she would have had a better and more proper reason to be proud of her country.

GIVEN THIS unfavorable atmosphere, I devoted my time in Aspen to anything but scientific interaction. I jogged a lot, did plenty of yoga, hiked up mountains, and played various sports. But while down in my office I gave myself an even more strenuous task, one that would occupy my mind during the time I spent there.

From the outset, Andy had been concerned that solving the horizon problem by no means implied that you had solved the homogeneity problem. You may have found a way of connecting the whole

---

*At least if one discounts the protests of an American organization called the Women Patriot Society, whose opinion of Einstein was that "not even Stalin himself was affiliated to so many anarchic-communist groups."

observable universe at some time in its past, thereby opening the doors for a physical mechanism to homogenize the vast regions we can see nowadays. But you still had to find that homogenizer, the mechanism pervading the baby universe that ensured it looked the same everywhere. In the language of science, solving the horizon problem was a necessary but not sufficient condition for solving the homogeneity problem.

Andy's wisdom was informed by experience. He happens to be a seasoned cosmologist, which means he has made a lot of errors in the past.* His seminal inflationary model suffered precisely from the drawback that although it solved the horizon problem it did not solve the homogeneity problem. Even though the whole observable universe was in contact during the period of inflation, when one computed what actually happened to its homogeneity one found a very irregular universe indeed. This was not a problem unique to inflation; indeed, Andy had told me that the bouncing universe suffers the same fate and that this had been Zeldovich's nemesis. Andy was afraid that VSL might fall into a similar trap, and during our meetings in his office he had voiced this concern persistently.

For the past months I had tried to ignore his questions in this respect because I knew that answering them meant performing a truly beastly calculation. If you want to see a cosmologist throw up, just mention the words "cosmological perturbation theory"; it's one of the most ghastly subjects in cosmology, and can cause even the best to break into a cold sweat.

We know that if we plug a homogeneous universe into the Einstein field equation, the Friedmann models naturally emerge. The idea is to repeat the calculation for a "perturbed" universe that in addition carries small density fluctuations around a uniform background. In some regions the density is slightly higher than normal, in others slightly

---

*Using this criterion, in precious few years I, too, will be a seasoned cosmologist.

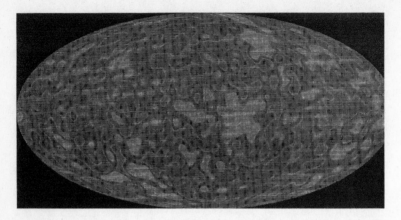

FIGURE 9.1   A picture of the cosmic radiation taken by the COBE satellite. The fluctuations in temperature are very small (about a part in 100,000) and represent the seeds for the formation of structures in our very homogeneous universe.

lower. You want to know if the "density contrast," as we call it, is suppressed or enhanced as the universe expands. To find that out, you insert your perturbed universe into the Einstein field equation and out comes a formula describing the dynamics of the fluctuations. It's a hell of a calculation, requiring endless pages of the most tedious algebra imaginable—the sort of thing a first-year Ph.D. student in cosmology does once and then spends the rest of his life trying to forget.

Complex as this calculation may be, the result is absolutely essential to the understanding of our universe. The cosmic radiation suffers from small ripples (see Figure 9.1); the galaxy fluid is homogeneous only on very large scales—on small scales it is made up of galaxies, which are not exactly uniform! Clearly in its fine detail the universe is not homogeneous, and this can be explained with that horrible piece of work, "cosmological perturbation theory."

To address Andy's quibbles and allay his concerns, I had to perform a similar calculation for VSL. A varying speed of light only added to the technical complexity of the problem. But I was sufficiently mentally bored in Aspen to have a go at it.

The first time I did the calculation it ran to some fifty pages of intricate algebra. I am not bad with lengthy calculations, but this one was so elaborate that I knew the chances of not having made an error somewhere were next to nil. Still, the final result was very pleasing. The result was a complicated differential equation that described the evolution of fluctuations away from homogeneity in a universe subject to VSL. When I solved it, the implication was that in addition to solving the horizon problem, VSL also solved the homogeneity problem. A big sigh of relief echoed throughout the valleys of Aspen.

Under VSL we could build the whole observable universe from a region sufficiently interconnected by fast interactions for thermal processes to make it uniform, in the same way that the temperature inside an oven is even because heat flows all over the place, homogenizing the temperature. But even the best oven is also plagued by temperature fluctuations arising because, as the heat currents flow, there is always a random chance that a given region is left hotter or colder. What I had just found, from my cosmological perturbation calculations, was that a varying speed of light would massively suppress such fluctuations. It just popped out of the algebra, and I didn't understand why, but the final result was that VSL predicted a totally homogeneous universe, one having no fluctuations whatsoever.

Hence we could not explain the structure of the universe or the ripples in the cosmic radiation, but we could set the stage for some other mechanism to come along and perturb the perfectly uniform background left over by the VSL epoch in the life of the universe. This news was as good as we could reasonably expect. After all, it took years before inflation was transformed from a solution to the riddles of the universe into a mechanism for structure formation, thus explaining the ripples in the cosmic radiation and the galaxy-clustering properties. We never thought that VSL would immediately appear to us in a final form capable of explaining all these features. The nightmare scenario for us, on the contrary, would have been to find that while we had solved the horizon problem, the universe still remained very inhomogeneous. My calculations excluded just such a possibility. Still, these were early days—should we blindly accept that pile of algebra?

I tried to persuade Andy to do the calculation independently, to see whether he obtained the same answer, but he refused outright, saying he was far too old for that kind of thing. Therefore, I decided to have another go at it myself. I waited a few days so that I would forget any possible errors I had made, and then started what I hoped would be an independent second calculation. This time I discovered a few tricks, shortcuts that reduced considerably the amount of algebra to be done, so that the second run filled a mere thirty pages. To my disappointment, the final equation was mathematically different. However, it still had the property of violently suppressing any density fluctuations, leading to a very homogeneous universe. Still, we now realized that, in detail, at least one of the results had to be wrong. A disappointed sigh now echoed around the mountains.

So it was with these humongous calculations, and their ups and downs, that I occupied my working time in Aspen, while a few yards away everyone else competed over some fine detail of inflation. It was a lonely and boring task, but it felt good to be so detached. Occasionally, I wondered what people around me would think if they knew what I was up to. That I was committing scientific suicide . . . That I was wasting my time . . . That I had gone mad . . . Funnily enough, one day I caught someone scrutinizing my calculations, left scattered on my desk. I had approached my office silently and no one was around. Through the door hinges I could see his sly face as he went through my pages of figures and formulae. I never let the "spy" know he'd been caught because it was just too funny a picture: He looked like a kid stealing sweets. Anyway, I was sure he wouldn't understand a word. What I was doing was so alien that he probably thought cosmologists in England use a special cipher to protect the privacy of their work. Such were the mental vibes floating around the place.

But I am probably painting too dark a picture of this meeting. If I considered my stay in Aspen as a vacation, it was fun and relaxed enough. I hiked with people, we saw videos over beers together, and we went out every night. I disliked Aspen's snobbery, but when we finally discovered a Hispanic nightclub out of town the real fun began. I also played a lot of sports, in particular soccer, in which my performance is uniformly laughable despite my national origins. Playing with scientists was fun enough—the Russians never passed the ball to anyone (including other Russians), the Latin Americans beat world records in fouling. . . .

One day we accepted a match against a group of local kids who obviously spent every waking moment in the gym. Euphoria hit the scientists when they beat the locals 10–0, a success partly due to Andy and I having teamed with the local boys to even up the numbers. While the scientists celebrated, the kids threw dirty looks at the two of us, clearly wondering whether we were some kind of Trojan horse—but I swear it was genuine incompetence.

WHEN I GOT BACK TO LONDON, I started at once making moves towards buying a flat. My Aspen trip had convinced me that I wanted to settle in London for a while. Up until that point, I had always vaguely considered the United States a possibility. I then went to see Kim in Swansea, South Wales, where she was a postdoc at the time.

"Swansea is the graveyard of all ambition." Or so goes the quote from Dylan Thomas, perhaps the only notable figure Swansea has ever produced. He both loved and hated the place, constantly running away, only to gravitate back to its seediness and low life. It says something about Swansea that none of its avenues, streets, or roads is named after him.

While at Aspen, some 3,000 meters high, I had done a lot of heavy exercise every day. In Swansea, at sea level, I found myself blood-doped. I recall embarrassing myself with my own energy at a circuits class while everyone else looked on, thinking I must be on drugs. This surplus of energy found an obvious outlet: Why not try to repeat that nasty cosmological perturbation calculation? Kim was lodging at the time in the house of a psychologist, and I locked myself in his study and decided to be done with it once and for all. At first I found it very distracting, as I started reading this guy's books and discovering more and more similarities between his behavior and the personality disorders he was supposed to understand. But after many happy hours reading psychology books, I finally became bored and managed to concentrate on the task at hand.

This time I found a fine trick that allowed me to do the calculation in three independent ways. None was too hard, and each one ran to some ten pages. But even better: They all agreed! Furthermore, they agreed with the answer I had arrived at the first time round, back in Aspen. I returned to London bringing Andy the good news—there could now be no doubt: VSL was definitely a solution to the homogeneity problem.

It was late one night, while I walked the London streets in the company of foxes, that it all became clear to me. I didn't need dozens of pages of algebra to understand the result. A simple argument would have been enough, what physicists call a "back of the envelope" calculation.

You may recall that VSL solves the flatness problem due to violations of energy conservation. A flat model must have a density at a given time (i.e., for a given expansion speed) equal to a crucial value, called the critical density. A closed model has a higher density, an open model a lower density. We had found that if the speed of light were to decrease then energy is destroyed in a closed, denser model, and created in an open, less dense universe. Hence you are pushed towards the critical density, that is, to a flat model, under VSL. This is what I called "the VSL valley of flatness."

I suddenly realized that it is precisely this process that enforces the homogeneity of the universe. Consider a flat universe with small

FIGURE 9.2 A density wave around a universe with the critical density. Overdense regions are like small closed universes, and so they lose energy if the speed of light decreases. Underdense regions are like small open universes, and so they gain energy. Either way the universe is pushed towards the critical density typical of a flat model. This phenomenon not only ensures flatness everywhere, but also produces a very homogeneous universe.

fluctuations around it. Regions where there is an over-density resemble a small closed universe because their density is higher than critical. Regions that are less dense must have a density smaller than the critical value, so they look like a small open universe. Now the equations describing violations of energy conservation are what physicists call *local;* that is, they only care about what happens in a given region rather than in the whole space. Hence energy would be destroyed in denser regions and created in sparser ones: You are still pushed toward the critical density everywhere. But this means suppressing density fluctuations, that is, enforcing homogeneity (fig. 9.2). In other words, the same argument that solves the flatness problem also solves the homogeneity problem. A little thinking should have been enough to arrive at this conclusion. How utterly stupid of me.

When I was a student at Lisbon University, I liked to play the smart ass—I would refuse to solve problems in the most straightforward way. For me that was almost as bad as getting them wrong. Instead, I would always try to find a clever way of doing them, not only getting the right result but finding it in a few lines instead of several pages. This occasionally infuriated some of my examiners. Becoming a real researcher has been the ultimate humbling experience for me. Nature is the examiner from hell; if you find new things at all, you always find them the hard way, with sweat and tears. Only then do you notice that there was a really easy way to find them. But this insight rarely arrives before you have been utterly humiliated and reduced to despair.

The good news is that you *do* find things out, one way or the other. This was driven home by a curious incident later that summer. After all those calculations I was ready for another vacation, and Kim and I spent a few weeks in Portugal, driving through the country in my dad's car, seeking out remote places at the end of the world. One day, we found ourselves miles away from civilization, at a remote sandy beach on the coast of Alentejo. As the sun set and we started to get the munchies, we prepared to go back to the world. And then Kim discovered that she had lost the car keys! The beach was huge and empty, there were few reference points, and the tide was coming in fast. With dismay, I started getting ready to spend a cold, hungry night outdoors, and to walk several miles the next day looking for help. But Kim didn't give up searching, even as it got darker and the water crept closer.

One hour later she found them. They were several inches under the sand, and in a few more minutes they would have been underwater too.

So whenever people tell me that finding a theory is like finding a needle in a haystack, I think of this incident. People do find lost keys in vast sandy beaches—sometimes.*

MEANWHILE, ANDY was growing colder all the time. Our meetings became shorter and more infrequent—he seemed to find them painful. I could sense his growing distance and discomfort. His gut reaction to anything related to VSL became very negative, not for the sake of improving the theory with constructive criticism, but from a need to place some distance between himself and our project. As a result, the writing of our article was dragging on forever. Andy kept

---

*Kim assures me that she *did* lose the keys, and was not pulling my leg the whole time.

finding new details yet to be worked out, and more excuses to post-pone submission. This went on and on throughout July and August. By the end of the summer, despite my recent victories, the whole thing seemed to be grinding to a halt.

I had a number of interpretations for this behavior. I've already mentioned that scientists often suffer from stage fright before sub-mitting new work for publication. One thing I now know is that you have to physically restrain the panicking author; if you let him have his way he will forever find excuses for delaying submission, and you'll never publish. It's a totally self-destructive pattern of behavior that can only be stopped by the other authors slapping the hysterical one in the face to restore order in the collaboration.

With something like VSL, so new and radical, I knew that we would never be completely sure we had it right. One just had to plunge in, and if the waters were shark-infested, well . . . too bad. In contrast, Andy's insecurities would simply drag the project into the wastebasket. I pointed this out to him, but at this stage I was too young to recognize how brutal you must be on such occasions. The crux of the issue is that it is always the author who has done less of the dirty work, that is, the most senior one, who gets into these pan-ics. It is perhaps some inner voice telling him that he should have done more. But the reaction is never to do some work, it's to express discomfort with the results. This is exasperating, and I began to regret having started this collaboration with Andy. Naturally, our rela-tions became strained.

But the definite break occurred when Andy turned forty in September, while we were at a meeting in St. Andrews, in Scotland. He invited a few of us to the house where he was staying with his family. I recall that Neil Turok and Tom Kibble were there. I myself had turned thirty a few weeks before, and we talked about the effects of age upon life in general and science in particular. Andy made a joke I will never forget. He said that now that he was turning forty he felt that the time was ripe for him to become a conservative and a fascist. At the stroke of midnight his personality would change, and the next day I would not be able to recognize him.

We all laughed politely, but as it turned out it wasn't a joke. It must have been more like a resolution. His personality did change noticeably overnight, at least as far as I was concerned. The next day, he announced to me that "these are very speculative ideas indeed, and not the sort of thing I want to have my name mixed up in." He stated that he was now the leader of the cosmology group at Imperial College, and that he could not afford having his image tainted by what he felt was just a bunch of cranky musings. He was supposed to give a talk on VSL in St. Andrews, but decided to talk about something else instead.

I was very shocked by this change of heart, but I should have seen it coming. Andy, in his "old age," was clearly trying to play the manager instead of the football player, a common move for scientists in middle age. As a manager, you show great interest in younger people's work, write the papers' blurbs, delay submission by requesting that more and more stuff be done, and finally put your name on all the publications. You then sit in science policy meetings that are sort of group psychotherapy sessions designed to give senior people the impression that they actually do some work.

This is a sad reality, and I could not believe that Andy was going down this path. There were a few people around us of a similar age who still did the dirty work side-by-side with their students, so age was not really the issue. Andy deserved better than that, and indeed, as I thought to myself, he was doing quite a poor impression of a manager. After all, he had managed me terribly. He had invited me to start this unconventional project, while steering me away from other more mainstream projects, and now had decided to scrap the whole thing. That amounted to a wasted year for me. Even from a managerial point of view, how would it look if I took my Royal Society fellowship elsewhere? Quite frankly, I started making plans to move, and if I did not take them further it was only because I really loved living in London.

I think Andy must have noticed that I was on the brink, and things did improve. The previous year I had supervised some of his students in a deal that gave him all the credit; he now gave me a proper Ph.D. student of my own, and he made sure I got the best one. I know he badly wanted this particular student, so it was a sacrifice for him. I felt he was clearly trying to make amends. He later apologized for the sharp words we had exchanged in St. Andrews, and ultimately he did not abandon the VSL ship. But his heart wasn't in it, and everything took ages. Again he apologized, saying he simply didn't have the time for it. After a slow and painful process, in November we finally submitted a paper for publication. And there starts another story: the fight to have VSL accepted by the wider community.

By December 1997, I was completely depressed. The last rays of pride and enthusiasm had vanished behind the mountains. I had spent a whole year working on a difficult project, which for all I knew was nothing but a pile of rubbish. From my perspective, VSL was confined to Andy and me, and all I got from him now was sheer rejection. In a world in which you are expected to publish four or five papers every year, I had not published even one. What had started off as a lot of fun had by now soured. It seemed to me that I'd just wasted one year of my life, and not even by being idle.

So when I went to the Jazz Café that New Year's Eve, I had good reason to share Courtney Pine's feelings. It had been a hard year, and I could only hope that the next one would be easier.

Of course, things can always get worse and, naturally enough, they did.

10   THE GUTENBERG BATTLE

SCIENTIFIC PUBLICATIONS are an important part of science and of a scientist's career. As an individual you are judged by how many papers you publish, where you publish them, their quality, and how often they are subsequently cited. But more importantly, publication is part and parcel of the fact that scientists, who tend to live on grant money, are obliged to make their findings and ideas available to others. They will not get their share of funding unless they can show a solid publication record.

Then there is the fact that before a paper is accepted for publication, it must undergo a peer review process. The editor of any respectable journal will choose an anonymous and hopefully independent referee, who is asked to study the paper and write a report on it. Based on this report the editor must then decide if the paper should be published or rejected, or if changes are required before the paper can be accepted for publication. Authors can in general reply to negative reports, and if in doubt, the editor may request expert advice from other referees or adjudicators.

There has been much debate over whether or not this quality control system works, but for the moment it is here to stay. It certainly leaves plenty of room for abuse. A pathological object lesson was the first "tentative" article on VSL, which Andy and I drafted at the end of the summer of 1997. We decided to submit it to *Nature*, a prestigious publication in which many major discoveries first appeared. These days this journal still carries forward its proud tradition in many fields, but unfortunately not in fundamental physics or cosmology, something which at the time I had not yet appreciated.

Unusually, *Nature* publishes in disciplines as disparate as biology and physics and each of them is supervised by a different editor. Hence I cannot comment on what goes on outside my subject. But in my field (although no one dares to say this publicly) there is consensus: They have employed a first-class moron as an editor. Some of my colleagues have now shown me many a report containing his judgments, but regrettably they will not allow me to publish a critical edition of these gems. Believe me, they are hilarious! The guy fancies himself to be a great technical expert and to prove it dishes out a stream of jargon, which is in fact pure gibberish.

Naturally you have to be a scientist to fully appreciate this, but fortunately his views on VSL were more down to earth; hence I am able to provide you with an example of the meanderings of this unique mind. Before submitting our article we sent to *Nature* a short summary of our work, detailing how a varying speed of light could solve the cosmological problems. Back came a report congratulating us on our efforts; however, we had to realize that our work could not be published in *Nature*. For that, we were told, we would have to do more than show that our theory was *a* solution to the cosmological problems—our theory would have to be *the* solution.

What can this possibly mean? How do you know that you hold *the* answer, not just *an* answer to a cosmological mystery? Does such a thing exist? And if this criterion were to be uniformly applied to all submissions, would any papers ever be published? Perhaps God's complete works will one day be accepted for publication in *Nature*, but even that is far from certain.

We are obviously talking about a failed scientist—envy of the penis springs to mind. It's a sad fact that the world is full of such peo-

ple: literary critics, art curators . . . people who possess both considerable power and frustrated bitterness.*

Needless to say, this article never saw the light of day (which may
well have contributed to Andy's middle age crisis), and instead Andy
and I decided to concentrate on a longer paper, containing as much
detail as possible. In November 1997, we finally submitted a technical account of VSL for publication in *Physical Review D* (*PRD* for
short), the same journal where some twenty years before Alan Guth
had published his theory of inflation. In general, my papers in *PRD*
had all been accepted within a few weeks of submission. "Time
Varying Speed of Light As a Solution to the Cosmological
Problems" was to drag on through a rather contorted year-long
review process.

Even within the tradition of robust debate that characterizes most
scientific arguments, the first referee report we received bordered on
the insulting. It stated that our approach was "unprofessional," even
though the report itself contained next to no scientific content in
refutation of our arguments. But if I found the report mildly offensive, Andy went through the roof. Innuendos in the report tipped
Andy off and he guessed the identity of this anonymous referee, one
of his arch-rivals from the early days of the development of inflation. I'm afraid that this is one of the drawbacks of the refereeing
process: All too often people use it to settle personal disputes.

That first report naturally led to a series of replies and counter-
replies, at the end of which everyone was accusing everyone else of
behaving irrationally. Our first reply included such gems as: "So far

---

*Of course, the problem is that the cosmology papers that do get published
in *Nature* are totally irrelevant. When I finally noted this anomaly, I stopped
submitting papers there, and proudly inscribed in my CV that none of my
papers has been published in *Nature*. But I do miss the light relief provided by
those reports.

the only thing 'unprofessional' about this business is the fact that [the] referee has worked himself into such an emotional state that he found it necessary to question our scientific professionalism. Both of the authors have well established reputations based on our strong records of achievement in the field. We have judged it worth our effort to put our reputations behind some interesting speculative ideas, and that should be the end of the question of 'professionalism.'"

It went downhill all the way from there.

By the end of April 1998, it was obvious that the refereeing process wasn't going anywhere. Other referees had by now been consulted, but the existing correspondence (which is always made available to new referees) was such that no one wanted to come down on either side, lest they get caught in the crossfire. Finally, the editor, in an act of heroic altruism, decided to step in and act as a referee himself. He happened to be an expert in the field, and stated his own misgivings about the idea. We didn't believe his criticisms were valid, but were happy to find that at last the fight seemed to be about science rather than scientists.

IN CASE YOU THINK that bitching is all there is to these refereeing battles, let me disabuse you by noting that there can frequently be up to 1 percent scientific substance in these reports. Indeed, amidst the barrage of insults, even that first referee report contained one lone piece of scientific argument. In a rare moment of calm, the referee pointed out that VSL lacked a "least action principle formulation." That was true, and at first it worried me too. Action principles first appeared as a beautiful reformulation of Newtonian mechanics, and nowadays they are the framework within which all new theories—with the exclusion of VSL—are proposed.

Newton's work *Principia* quickly became the bible of physics, but not everyone felt comfortable with its philosophical implications.

Newton's view of the world is unashamedly deterministic and causal-
istic. It contains a system of equations stating that if you know what
every particle in the universe is doing at a given moment, then you
can predict their futures exactly. It's a formalism aimed at linking
cause and effect in a perfect mechanical chain where deviations are
not allowed. Taken at face value, this has always greatly annoyed "free
thinkers."

In Newton's world, everything happens for a reason, that is, by
virtue of some cause. But precisely for this reason the Newtonian
clockwork universe is devoid of meaning in the human sense of the
word. God acted upon the world when He created its laws of causa-
tion—but then left the world to its own devices. The Newtonian
world has as much meaning and purpose as a clockwork doll, as
opposed to an act of love. The problem is that one may then claim
that an act of love is also ruled by Newton's laws—a most displeas-
ing thought.

In 1746, the French physicist Pierre de Maupertuis found an alternative way to describe the physical world. He considered the trajectories followed by particles in mechanical systems, and he noticed a
pattern: It was as if particles, moving along their paths, were trying
to minimize a certain mathematical quantity, which Maupertuis
dubbed "the action." He thus was able to reformulate mechanics by
stating that nature behaves in a way that minimizes the action—the
so-called least action principle. This was exactly the kind of formulation Andy and I had been unable to find for VSL.

This approach may sound alien to you, but believe me, it is mathematically equivalent to the Newtonian framework. But this was not
fully appreciated initially, or at least people got tangled up in games
of words, mixing physics with philosophy and religion in an inglori-
ous salad not uncommon in those days. In Maupertuis's world, it
looked as if there was finality instead of causality: Things happened
toward an end (that of minimizing the action) rather than *due to a cause*.

Maupertuis's world—unlike Newton's—had purpose or intent. Take one little step further and you'll be proving the presence of God in the everyday workings of nature, rather than just at the moment of its creation. For God, being naturally lazy, would minimize the "action" in His deeds.

Nowadays, this feels very far-fetched, but it reflects a common trend in the philosophy of the time: Leibnitz's optimistic doctrine, according to which we live in the best of all possible worlds, by the grace of God. Maupertuis's mechanics was certainly optimized, with minimal waste of action, so it would seem that Leibnitz's philosophy had a scientific basis. However, so similar were these ideas that Maupertuis soon found himself involved in a nasty priority dispute over the principle of least action. But even worse, he also acquired Leibnitz's enemies, in particular his most vocal opponent, Voltaire. It is on this account that a major altercation would be recorded in the annals of physics. Despite all the parallels, the VSL publication battle pales by comparison.

You may have come across a novel by Voltaire, *Candide,* in which a naïve young man lives through the chaos and sufferings of a brutal world, always maintaining that his trials and tribulations are for the best, in the best of all possible worlds. It is a cruel parody of Leibnitz's philosophy, which to this day can cause convulsive laughter. In fact, Voltaire was a playboy and an incorrigible satirist, but also a philosopher who firmly believed in God as a clockwork designer, and in His direct absence in the day-to-day running of nature. In his usual style, Voltaire stressed how most of the devastation caused by the 1755 Lisbon earthquake was due to its unfortunate timing: on Sunday morning, while everyone was at mass— meaning that plenty of candles were lit, ready to unleash a raging fire.

Voltaire despised Leibnitz's philosophy and unsurprisingly also turned his guns on Maupertuis and his least action principle. That Voltaire and Maupertuis had had an affair with the same woman in a complex menage à quatre (if you also count her husband) may also have played a role in this "scientific" dispute.* Whatever the case, in a pamphlet titled "The Diatribe of Dr. Akakia," Voltaire portrayed Maupertuis as a paranoid and lunatic scientist who vivisected toads to learn about geometry, advocated the centrifugal force as a cure to apoplexy, performed trephinations in humans to penetrate the secrets of the soul, proved the existence of God by Z equals BC divided by A+B, and much, much more along similar lines. Sadly enough, all this nonsense is obliquely based on pieces of research actually carried out by Maupertuis.

So serious are Maupertuis's derangements that Dr. Akakia, an expert in psychological diseases (and surgeon of the pope!), is called upon to provide emergency treatment. Dr. Akakia finds his demented patient so far gone that he approaches the Holy Inquisition seeking excommunication as a form of psychotherapy; in return, his patient, invoking the principle of least action, attempts to kill him.

Voltaire's essays on Maupertuis have become a nasty monument to the power of caustic sarcasm. For months, high society did nothing but laugh at Maupertuis's expense while quoting from Voltaire's pamphlets and firmly ostracizing the poor man. Maupertuis became the laughing stock of Europe and in despair sought refuge in Switzerland. His health never fully recovered and he eventually died—some claim—of shame.

So here are the gutters of science, displayed for your benefit, with all they have in common, a few centuries apart. There have always been and always will be some scientists who feel personal insult to be far more satisfying than a rational argument. Today we know that Maupertuis was a much better scientist than Voltaire, but he lacked Voltaire's resourcefulness with words and philosophy. The latter, however, was far easier for the public to understand.

Another aspect of Maupertuis's story that is relevant to the VSL battle concerns the refereeing system of his day: the Inquisition.

---

*To the best of my knowledge, this is one aspect of Maupertuis's story that bears no relation to the VSL publication battle.

Indeed, many of Voltaire's works (including the "Diatribe") were burnt. Nowadays, papers are no longer burnt as heretical,* but some things are still the same. I have in mind Voltaire's views in "Micromegas," the tale of an inhabitant of a planet orbiting Sirius. In his youth, the hero of the story writes a most interesting book on insects; however, "the local mufti, who was a great pedant and extremely ignorant, found some of the arguments in his book to be suspect, offensive, foolhardy, and heretical, indeed steeped in heresy; and he proceeded energetically against it. The case turned on whether the substantial form of the fleas on Sirius was of the same nature as that of the snails. Micromegas defended himself with wit and won the ladies over to his side. The trial lasted two hundred and twenty years. In the end, the mufti had the book condemned by legal experts who had not read it, and the author was ordered not to appear in court for the next eight hundred years. He was only moderately grieved to be banished from a court full of nothing but needless fuss and pettiness." A stream of analogies spring to my mind.

But let's leave aside the omnipresent scientific manure and examine the accidental morsel of wisdom provided by that first referee. Why did Andy and I not formulate VSL via an action principle?

It is clear that VSL contradicts the theory of special relativity, which is based upon two postulates: the principle of relativity (i.e., the statement that movement is relative) and the constancy of the speed of light. Combining these two principles leads to a set of laws, called Lorentz transformations, that describe how to relate the world as seen by different observers moving with respect to one another. Lorentz transformations encapsulate how time dilates and distances shrink, and a theory in which all quantities obey these transformation laws is

---

*Instead, as publication of the British edition of this book would soon reveal, controversial books are now pulped (which besides being less polluting is so much more politically correct). The author would like to thank those keen to act as modern-day Inquisitors—such as the "scientific" journal *Nature*—for no longer playing with fire. (Note added to the U.S. paperback edition.)

said to satisfy "Lorentz symmetry," or be "Lorentz invariant." In such a theory all laws symmetrically mimic the Lorentz transformations.

Lorentz symmetry, apart from its physical significance, is a mathematical asset: It makes many equations and laws much simpler to write down. In particular, the mathematics of action principles loves Lorentz symmetry, although it does not specifically require it (after all, action principles were discovered in the eighteenth century, long before relativity). Action principles seem to fit like a glove on Lorentz invariant theories.

VSL naturally conflicted with Lorentz symmetry because it was based on demolishing its second main principle, the constancy of the speed of light. Therefore an action principle formulation of VSL was extremely awkward, and it was not until much later that I worked out how to do it. But was this an inconsistency?

Of course not! It is a very recent trend to write down a new theory by means of an action. Relativity itself was not formulated via an action at first, even though actions are particularly suited to it. Despite all the seemingly different philosophical implications, it is a matter of convenience whether you formulate a new theory using the language of Newton or Maupertuis. VSL seemed to be more pleased with the former; so what?

Now, if you will, picture having a scientific argument on this issue with a referee behaving as if he had been bitten by a rabid dog.

WHILE THIS IMPROVING battle of insults was taking place, two things happened. First, I managed to convince Andy that the refereeing process was taking so long that we should distribute copies of our manuscript to a limited number of people. One of these was John Barrow, a scientist with a strong record in so-called "varying constant" theories. John was immediately fascinated by the concept, and he started asking us a lot of questions about our paper.

Andy was very worried by this. I recall Andy telling me: "Look, João, suppose he goes off and writes a paper on his own containing

JOHN BARROW

no reference to our work. He then sends it off to *PRD* and is lucky enough to get a referee more intelligent than the idiot we got—after all, it's a bit of a lottery. What could we do? I don't know about John Barrow, but in the U.S. this could very well happen and I tell you more. If you then go around and complain to other people about what happened, they would just laugh at you for being so stupid."

I thought this was a bit of an exaggeration, but I did ask a friend of mine, who had previously worked with John, what he thought about the situation. He answered, "I may be very wrong and he may turn out to be a total arsehole, but from my experience John is the most reliable guy I have come across."

A few days later, we received the news that John was indeed writing a paper on VSL. Spectacularly, a couple of weeks after receiving our paper, John had his own version of VSL written up and submitted for publication in *PRD*!

Naturally, there was consternation in the Albrecht-Magueijo ranks, the more so because I then left for Australia, and due to various short circuits we didn't see John's paper for quite a while. In the middle of this predicament, I recall thinking to myself that the only way out might be to arrange for John to collaborate with us—a form of damage control, to be sure, but better than nothing. It looked as if we were going to be scooped and that Andy's worst fears would be realized.

My trip to Australia, however, would place my mind on a very different plane as I absorbed the "no worries" cliché with the greatest pleasure. Australia is Kim's homeland, but she had been away for more than six years; so we took this opportunity to stage a road-movie scenario, driving over 7,000 km in a few weeks. The trip was a lot of fun, and while insulting e-mails flew back and forth, and the danger of being scooped loomed over us, I nevertheless had a very relaxing time in a land I really loved. It was the perfect therapy.

Like a cosmological model once proposed by the famous physicist Milne, Australia has more space than substance—which is precisely what makes it so attractive. Most of it is either desert or lush jungle inhabited mainly by crocodiles. Occupying an area not much smaller than the United States, Australia still manages to have a population not much larger than Portugal's. The animals are still in charge, to the great horror of many a European visitor and the ensuing amusement of local onlookers.

For hours and hours we drove through nothingness, a philosophical contradiction bound to put you in a strange frame of mind. As the roads unfolded, occasionally (sometimes *very* occasionally) in the middle of nowhere we would cross some tiny God-forsaken town with a name like Woolaroomellaroobellaroo, and home to only a few dozens of souls—but always blessed by Napoleonic town planning: huge pavements, imposing boulevards, wide avenues full of nothing. Clearly, the welfare state had been at work; indeed Australia feels like a cross between Denmark and the United States: a welfare state—but with hormones.

Other times, we would go for a whole day without seeing civilization, just sporadic empty watercourses, invariably carrying an imaginative name: 2-mile creek, 9-mile creek, 7-mile creek, 3-mile creek, etc., etc., etc. My mathematical mind found itself building a histogram capturing the distribution of lengths of Australian creeks. As I said before, in such emptiness, your mind flutters away into surreal pastures.

But I'd decided to do more than just be a tourist, so in addition I gave talks at several universities. I liked the people I met very much, in particular their blunt views on trends in cosmology. As a random recollection, in Melbourne I met Ray Volkas, who upon hearing my idea simply said that VSL was no more or less dodgy than inflation and was at least more interesting. I also met Paul Davies in Adelaide, who a few years before had left his university job to devote himself to writing popular science. For that decision he has been crucified by other scientists, but I gave him points for not having become a bureaucrat like most of his critics. Besides, while we walked down the campus streets, I couldn't help noticing that all the pretty girls nodded to him.

Then in Canberra I met a group of astronomers working at Mt. Stromlo, an observatory surrounded by a large population of kangaroos. This was only the second time I had gone anywhere near a telescope, the first being when I was an undergraduate and gave a hand to someone doing an astronomy project. On that occasion I recall dropping the dome door on top of the telescope mirror, causing an explosion of swearing—but miraculously not breaking the mirror itself. Still, my help was not enlisted again. Now, in the middle of a sea of kangaroos, I reflected on how far astronomy had come since Hubble's days, with ever-improving technology and sharper and sharper data, forcing cosmologists to look hard at the real world before engaging in their flights of fantasy. And predictably enough, the Mt. Stromlo astronomers sneered at VSL as a figment of my imagination.

But the real action would take place while I was visiting the University of New South Wales in Sydney.* At the time, John Barrow

---

*Unlike the UNSW, and despite its lack of scientific output, the physics department of the University of Sydney is too self-important to accept visits from mere cosmologists. In this light it would not be impertinent to suggest that they should give a professorial job to the aforementioned editor of *Nature*.

was the director of the astronomy department at Sussex University, just down the road from London, but we had never met. We happened to meet for the first time in Sydney during that trip, but our first meeting was a total disaster.

John had just given a public talk (entitled "Is the World Simple or Complicated?") with his usual brilliance. I recall that there was a four-year-old girl in the audience, and such was John's clarity that she had listened attentively and even asked a pertinent question at the end.

Afterwards John Webb, our host, took us to dinner at a nice restaurant by the pier, and that's where a huge argument broke out. John Barrow and I are on opposite ends of the political spectrum, and over dinner his conservative leanings led him to make some inexcusable statements. Kim and I ended up shouting at him, with the odd cavalry charge from Webb's wife thrown in for good measure, while people at other tables kept staring at us. Even in Australia it is not usual to start a riot in a posh restaurant.

After that, I thought it better to forget about collaborating with John. The following day, however, we met at the university and started talking science. At once there was perfect mutual understanding, and over the next year we would write four papers on VSL together. I have always been fascinated by science's ability to bring together people who are otherwise completely incompatible.

It was also in Sydney that I finally saw a copy of John's paper on VSL. Andy and I could not have been more misguided in our worries. John was very careful about giving us credit, and even referred to VSL as the "Albrecht-Magueijo" model. His speedy composition of a paper on VSL reflected genuine enthusiasm rather than an attempt to scoop. For me, this suggested that interest in VSL might soon spread to a sizable portion of the scientific community, which pleased me a great deal.

But I was to learn of another, far more exciting development during this visit. A group of Australian astronomers, led by John Webb, had discovered evidence for what just might be a varying speed of light. This was exciting news! I even thought of driving all the way back to Mt. Stromlo to rub it in their faces. Of course the result was itself controversial, and it still could be interpreted in other ways, but at last it looked as if our theory might actually be better than inflation in one crucial aspect: There might be *direct* observational evidence for it.

As I said before, the speed of light, the "c" in the equations, is woven into the very fabric of physics, and has implications well beyond cosmology. It turns up in the most surprising places, for example in the equations that govern the way electrons move inside atoms. In particular, something called the "atomic fine structure constant" (often abbreviated as "alpha") depends upon c.

When light shines through a gas cloud, its electrons absorb light of certain specific colors, producing a pattern of dark lines in the spectrum, reflecting the steps of the energy ladder on which the electrons live inside atoms. Upon closer inspection, however, one realizes that some of these lines are in fact several lines very close together—the atomic spectra have a "fine structure." The "fine" pattern depends on a number (naturally enough) called the fine structure constant, and laboratory scientists have used this fact to estimate the value of this constant with considerable precision. Perhaps unsurprisingly, c appears in the mathematical expression for alpha, so that looking at light spectra is a way of measuring the speed of light.

Interestingly, the same trick can be performed by astronomers with even greater accuracy by looking at light shining through distant clouds. The work done by John Webb and his team showed that while light from nearby galaxies confirmed the laboratory values for alpha, light from faraway clouds seemed to suggest that the constant was different. Now, when we look at faraway objects we are also looking at them in the past, since light takes time to cover the distance between them and us. Webb's results seemed to suggest that alpha was changing over time. If he's correct, one possible explanation (and I'll discuss alternatives soon) is that c is getting smaller! These results have yet to be confirmed, but they are extremely suggestive, and as such can be counted a triumph of VSL theory. The greatest compliment to any theory can only be paid by nature—when it is found that the theory predicts correct experimental results.

I returned from Australia in an excellent mood, carrying with me three major new assets. One was a new collaborator, where before had hovered the specter of competition; another was a striking piece of suggestive evidence. But back in London, most people only noted the impressive suntan.

THE MONTHS THAT FOLLOWED were spent performing what John Barrow would later refer to as "the reeducation of the *PRD* editor." It was an arduous process, but at least once the editor stepped in, the battle shifted to purely scientific grounds. Some of the questions raised were completely beside the point, but others were spot-on. My diary entries during this period constantly refer to the importance of learning how to take criticism. If you shut yourself in your own little world, that will be the death of your theory. On the other hand, many of the criticisms you receive are pointless and simply reflect the view that anything new is bad. In such a delicate situation it is crucial to tread gingerly and be careful to appreciate the difference between pertinent and idiotic comments.

Becoming dogmatic with something like VSL is like shooting yourself in the foot. I have since met several dogmatic physicists, ignored by all and carrying huge chips on their shoulders. I have also noticed that in their old age they become stone deaf. The Lamarckian theory, whereby disused organs are phased out, may play a role in this phenomenon.

But let me give you an example of the kind of issue that was at stake in these arguments. A major objection raised by the editor concerned the physical meaning of measuring a varying c. Surely you can always define your units of space and time so that c does not vary, wrote the editor. I was very puzzled by this comment because it is obviously true. Suppose someone tells you that the speed of light was twice its current value when the universe was half its present age. Because you don't like that, you decide to regauge all clocks used when the universe was half its current age so that they tick twice as fast. And presto . . . the speed of light becomes the same.

Andy and I argued at length over this and we soon saw that there had to be a flaw in the argument. After all, you could also regauge your clocks so that the speed of light becomes variable even in circumstances in which it is normally taken to be a constant. A simple way to do this would be to stupidly take a grandfather clock on a space mission. On the Moon, a pendulum clock ticks more slowly (since gravity is weaker), and if you insist that grandfather clocks are the correct way to keep time, then naturally you would find that the speed of light on the Moon is much higher.

There had to be a loophole somewhere. I thought about this . . . and thought . . . and got into a terrible tangle. No matter how I turned the problem around, I couldn't see how to evade the clever logic of the *PRD* editor. Eventually, I realized where I had to look for inspiration: John Webb's observations (which, incidentally, were unknown to the *PRD* editor). There I had an example of an experiment that could be interpreted as a varying c. Was it a fallacy? Had John Webb unwittingly employed a grandfather clock in his observations of the early universe?

Upon closer inspection the answer was a spectacular no. The fine structure constant alpha is given by the ratio of the square of the electron charge (e^2) on one hand, and the product of the speed of light (c) and Planck's constant (h), on the other. If you work it out you'll find that both sets of quantities are measured in the same type of units: energy times a length. But because the fine structure constant is a ratio of two things measured in the same units, it does not itself have units.

This is true in precisely the same way that pi (i.e., the number 3.14 . . . you learned in school) does not have units. Pi is the ratio of the length of a circle and the length of its diameter, that is, two lengths. Pi therefore takes the same value whether you measure your lengths in meters or in feet. Likewise, alpha is a pure number and its value does not depend on the units or whether you measure time using an electronic watch or a grandfather clock. Thus the issue of the constancy or variability of the fine structure constant, as addressed by John Webb and collaborators, was beyond the criticism raised by the editor. No matter what games you played regauging clocks and redefining units, you would always see a varying alpha.

But now you have a problem. If John Webb had found that alpha was indeed a constant, we would all be happy to say that e, c, and h are also all constants. But he didn't: He found that alpha changes over time, so which of e, c, or h do you blame? The situation is tricky. Whichever answer you give to this question, you are assigning variability to a constant that does have units, so you fall prey to the criticism of the *PRD* editor, that is, you can always change the units so that your chosen "varying constant" actually becomes constant. But you have no other choice; stating that none of them varies is not an option. So which of e, c, and h *is* varying?

Andy and I quickly realized that simplicity alone guides you to the answer. Your choice amounts to the specification of a system of units, which is of course arbitrary. But in practice, there is always one system of units that makes life simpler, for instance, measuring your age in seconds or in years is a matter of choice, but if I told you that I am 1,072,224,579 seconds old you might think me a little odd. Likewise, any choice of units is dictated by the simplicity of your description, and this choice determines which dimensional constants are *assumed* to vary.

VSL was a theory of nature in which alpha varied in such a way that the simplest way to describe it was to choose units so that c (and possibly also e or h) varied. To make this clear, John Barrow and I performed an interesting exercise in which we mathematically changed the units in our VSL theory so that c became a constant. The result was such a long-winded mathematical mess that we believe our point was taken. A changing c, as the editor had pointed out, *was* a choice or convention. But it was the right convention to adopt in the context of a theory based on contradicting relativity, as ours was.

In our theory, relativity was in tatters; the principles of Lorentz symmetry were not valid, the invariance of laws in time was lost, and a whole host of other novelties had been introduced and were part of the predictions of the theory. It made all the sense in the world that if we abandoned one of the pillars of Lorentz invariance, the constancy of the speed of light, we should use units bringing that fact to light; and that if we did so, the result would be a more transparent rendition of our theory.*

Funnily enough, this discussion with the editor reminded me of a major frustration I had encountered while teaching myself physics and mathematics, struggling to understand *The Meaning of Relativity.* I recall becoming exasperated with the way most physics books continually made use of the results they were trying to prove. Take the principle of inertia, which states that if not acted upon by a force, particles maintain a constant speed. But what is a constant speed? To measure a speed, you need a clock. How do you build a clock? And here starts the trouble: Books are either evasive on this matter or shamelessly use the physics they are trying to prove (such as the law of inertia) to construct a clock. The whole argument seems hopelessly circular.

I got so fed up with this that I decided to set the record straight and write a physics book myself. This proved to be a disaster because no matter how I tried to reformulate mechanics so that nothing was circular, not one of my attempts was watertight. Statements like the law of inertia always ended up in tautologies, and I would start all over again.

---

*John Barrow and I also found *other* theories in which it is best to blame the electron's charge for the variations in alpha. They were very different from VSL, and made very different experimental predictions.

But wait. The constant speed referred to in the law of inertia and the constant speed of light postulated in relativity have something in common—they are both speeds, after all. And finally, after this argument with the referee, I understood why I had failed in my youthful attempts to set physics straight.

Most propositions in physics, such as the law of inertia, the uniformity of time, or a varying speed of light, *are* to some extent circular, and amount to nothing but a definition of a system of units. The law of inertia tells you nothing more than that there exists a clock and a rod by means of which the law of inertia becomes true. It does not force you to use them and it does not make a statement that could ever be proved—noncircularly—by experiment. The law of inertia merely tells you that if you use this clock and rod your life will be easier. You can then formulate Newton's laws in a straightforward way, and eventually you do glean from the whole construction some propositions that are not circular, and that give you predictive power.

It is unavoidable that some aspects of physics are tautological or mere definitions; but the tautologies are never gratuitous, and the whole theory always includes a modicum of propositions with actual meaning. Hopefully the definitions you introduce make the real content of the theory clearer.

We included a new section in the paper spelling out this point of view and the editor withdrew his criticism. This was one of many occasions in which he had a valid point, but one we were able to address within our VSL theory. We continued to argue over the details of the theory in this way for another six months, and our manuscript grew to about twice its original size. Overall, there were more than seven rounds of referee report and author reply.

In hindsight, I have to admit that the paper was improved enormously over the course of this grueling exercise. By the end of the summer of 1998, things seemed to be converging. Slowly but steadily.

IN SPITE OF ALL the progress that had been made, there were still vicious moments. At one point, the editor of *PRD* visited Imperial College, and let's just say that what began as a polite scientific argument rapidly deteriorated into mayhem. In an attempt to redeem ourselves, Andy and I walked the poor man to the tube station on a glorious sunny day; but not many words were exchanged by this stage—the editor of *PRD* was sulking.

Once when the *PRD* editor took several months to produce a reply, I suggested submitting our paper to another journal simultaneously (a very illegal procedure), arguing that if we were being messed with we were entitled to return in kind. Andy, however, cut short this initiative, stating that the crucial thing in these fights is not to send all to hell and ostracize yourself.

In Andy's own enlightened words: "This bitterness thing is a vicious cycle. Anyone who appears to be an insider probably has had many experiences that might have prompted a retreat into bitterness. Reacting in a constructive way to these situations is what keeps insiders 'inside.'" Andy may have often played the "bad cop" in our refereeing battles, but unlike me he knew when to stop. I am immensely grateful to him.

In the last stretch of our long publication battle, as the summer warmth enveloped us, Andy's full enthusiasm for VSL returned and he even contributed calculations to our growing paper. Perhaps it was the adrenaline generated by the refereeing fight that gave him a kick. Whatever it was, our early sunny days came back in their full glory as our paper grew steadily and our ideas matured. For the sake of honesty and completeness, I've felt compelled to describe our dark season; but I would like to stress that many years on Andy and I have remained the best of friends. Perhaps this kind of love-hate relationship is the necessary crucible for all truly innovative ideas.

But even as this last stretch unfolded, and Andy and I once more began to click, a huge setback was in store: That summer, Andy would abandon Britain for an American university. Looking back,

I can see that all manner of pressures must have been taking their toll on him, until he finally got an offer he could not refuse. This was a major loss for British cosmology, but what really infuriated me was that Andy absolutely loved Imperial College. And still he left.

Britain has a unique ability to let its talent go. People like to say that it's because its academic institutions cannot financially compete with the United States, but I find that a poor excuse. In fact the British "brain drain" is totally self-inflicted, the product of a culture in which accountants, lawyers, consultants, politicians, and financial morons of all varieties are prized well above teachers, doctors, nurses, etc. It's considered bad taste in Britain to do anything useful these days.

But perhaps I should explain myself more clearly. Imperial College, and Andy knew this well, has perhaps the best scientific environment in the world. It has the best all-round students I have come across: clever, keen, and really fun to interact with. In a few other places students may be marginally better academically, but then they have nothing but academia to fill their lives. Students at Imperial, with their broader talents, are far more interesting.

Imperial also has an outstanding collection of researchers, both as itinerants passing through and as permanent tenured staff. When it comes to research, Imperial is a unique melting pot, partly because of its eclectic tendencies: its willingness to mix fields that are generally seen as incompatible (such as string theory and other approaches to quantum gravity, or inflation and cosmic strings).

Given the above, what more could Andy want? Well, quite a lot: Imperial suffers from endemic bad leadership. Its administrators always seem to be the last ones to realize that someone is doing well. And even when they do finally reward achievement, it is made to feel like a favor, and a considerable amount of muscle-flexing and humiliation is always mixed in. Unsurprisingly, successful researchers eventually feel unhappy and apply elsewhere, say in the United States, and get offers. Then all of a sudden Imperial's luminaries realize they cannot match these offers and start complaining about American imperialistic tendencies, when in fact, if they'd kept people happy in the first place they would never have applied elsewhere. Imperial's leaders are always one step behind and I would argue that they lack brains more urgently than money.

To be brutal, Imperial's leaders tend to fancy themselves as scientific pimps, in a scenario where the scientists are forced to play the whores. This descriptive wording comes from someone who left, and sums up the mood behind many losses. A few years before, Imperial had lost Neil Turok; that summer it lost Andy; and even as I write, the same mistake is being made with yet another scientist, this time a world-class string theorist. The people who want to cash in all the credit for a number one institution are, in my opinion, number one only for shit.*

But let me not be too harsh on them. These politicians of science are only following the edifying example of other administrators and politicians in this illustrious kingdom. Rather than rewarding their foot soldiers (read, those who actually do something), these people appear to be obsessed with their navels—spending their time fabricating statistics, leading huge administrative exercises aimed at promoting "accountability," and interfering with people's lives in matters in which they are not competent to provide advice.

For instance, recently we all had to report on what we did minute by minute over an entire week. This sort of thing is enormously disruptive, but more to the point, who cares about the statistics resulting from this type of expensive and time-consuming exercise?†

Another example close to my heart is the TQA, or Teaching Quality Assessment. The TQA supposedly brings accountability to

---

*To be fair, at the time of writing, I must be just about the only active researcher Imperial's wise men have managed not to alienate. Undoubtedly, I have been overlooked.

†In my reply, I included a very graphic description of all my trips to the toilet. No one complained, which makes me think that no one actually looks at these "exercises."

those who teach at British universities, thereby giving the government the impression they are doing something about education. But then we run into an unpleasant difficulty: How do you measure good teaching? Even worse, how do you measure it in a way that civil servants' brains can cope with?

Given the inherently subjective nature of the matter, the relevant officials struck upon a brilliant idea. Why not just measure the quality of your paperwork? That's objective enough—you get points for producing documents stating your aims and points for producing documents demonstrating that they were fulfilled. And no one cares that the system favors institutions that don't aim high. The lower your expectations the easier they are to fulfill.

The TQA literally generates tons of documents, mostly fake. Ironically, to produce them people have to take time away from lecture preparation and therefore stop teaching properly. All this is finally assessed by a bunch of bureaucrats and professors in third-class universities with a grudge against successful higher education. By the time the TQA is over, enough money has been spent to have kept a few dozen Andys in Britain. Furthermore, teaching quality has been considerably reduced. But the government is happy—universities are now accountable. It is only the civil servants who come up with all this shit who don't seem to be accountable.*

I wish this were a problem confined to higher education, but it's not. Schoolteachers have to *prove* that they have "added value" to their pupils, and to do so they have to stop preparing lessons and instead spend hours working with expensive government-issued statistical software, churning out meaningless numbers for the benefit

---

*I have been told that the TQA is part of the English class trauma that foreigners cannot understand. As alien as this might be to me, it's part of the government's strategy to give the working class the impression they are actually middle class, in this case by giving ex-"polytechnics" the feeling that they are proper universities. Or so I'm told by my British colleagues, who of course would never admit to this in public.

of government officials who will never set foot in a classroom and who get paid considerably more than a teacher. Meanwhile, it's becoming impossible to find anyone who wants to be a teacher in central London. Or a nurse. Or indeed anything useful. Being a parasite is so much easier and so much better-paid these days.

So even though I was foaming at the mouth with rage when Andy left, and seriously contemplating physical violence against the Master Pimp, I have to admit that seen from this wider perspective, Andy's departure—and the loss to British cosmology—was the least of our problems.

AS THE WINTER of 1998 set in, nearly four years after that gloomy day in Cambridge when I had my first glimpse into this new theory, VSL gained a modicum of scientific respectability as a slew of papers on the topic were finally accepted for publication.

On the one hand, my original paper with Andy, still growing in size, was getting closer and closer to publication but had still not been officially accepted; whereas my first paper with John Barrow, written nearly one year later, was accepted within weeks of submission with a very positive report (it *is* a lottery). John Webb's experimental results were also undergoing refereeing, and John Barrow was on the authors' list. No doubt all this was instrumental in triggering a wave of acceptances that finally covered all submissions in the area, including the Albrecht-Magueijo opus. The publication battle had been won.

With our baby finally in press, we decided to make our papers and ideas public at last. The first thing we did was place our papers on the Web in archives routinely read by physicists. Then *PRD* itself published a prerelease note.

I was not ready for what happened next. For all these years I had been readying myself for the possibility that my own passion for VSL might never spread to the rest of the scientific community, let alone the rest of the world. So it came as a complete surprise to me when the idea caught the fancy of the popular press who follow scientific publications. Brief newspaper articles were followed by yet more newspaper and magazine articles. Then I started to get requests for popular talks and radio appearances, and finally a documentary on VSL was commissioned by Channel 4 (a somewhat high-brow British channel). People were interested not only in the idea itself but in the origins of the idea—how I came to think about VSL as an alternative to inflation theory.

But almost as soon as I started to bask in the glory of acceptance, a tremendous altercation broke out. Imagine my shock when I discovered that another physicist had been there before us.

As we landed, a flag already flew over the Moon.

11   THE MORNING AFTER

IN 1992, JOHN MOFFAT, a theoretical physicist at the University of Toronto, discovered VSL as an alternative to inflation. His theory was totally different in form from ours, but the substance was very similar. That there could be *other* VSL theories did not surprise me: I knew from the beginning that VSL, like inflation, could come in many different flavors, and that we had merely chosen one for starters. What shocked me was that someone had toyed with the idea of a varying c before us, but the wider community hadn't noticed.

Moffat had written a paper describing his findings and submitted it to *PRD*. He was met with a reaction much like the one we would encounter a few years later. The outcome for him was entirely different, however; following a one-year fight with the editor and referees, Moffat conceded defeat! His paper eventually made its way into a minor journal I had never heard of. This was why Andy, John, and I had not been aware of Moffat's work.*

Now Moffat noticed "with chagrin" how our papers, containing essentially the same idea, were being accepted for publication in the same journal that had rejected his work. With a very hurt tone he wrote us an e-mail drawing our attention to his paper and requesting a citation. In addition, he approached *PRD.* He wanted them to stop

---

*Moffat also put his paper on the Web archives I mentioned earlier, but this was at a time when none of us used them routinely.

JOHN MOFFAT

publication of our work, and he even implied legal action over copyright issues. He was understandably furious, and one of his former Ph.D. students, Neil Cornish, whom I knew quite well, sent me an e-mail putting the whole episode in perspective:

> When he wrote the paper it was met with deafening silence. . . . Janna [Levin] and I were encouraging but Dick Bond* wasn't interested at all. [Moffat] sees Albrecht and Barrow as members of the establishment like Bond, so he must be thinking: "The establishment wouldn't take me seriously, but now they will take my work." I'm not saying that is what has happened, but it is the way he must be seeing it. I'll contact Moffat and see if I can help to calm him down some more. What are you and Andy thinking of doing?

I knew exactly what I was going to do: apologize profusely to Moffat and embrace him as a friend. I certainly felt he had every rea-

---

*The head of the Canadian Institute for Theoretical Astrophysics, and a staunch defender of inflation.

son in the world to be fed up with scientific journals. If I were thirty years older and had lost my publication battle, I would certainly feel the same way. Also, since our paper was still in proof, we could easily add a note explaining the situation.

Of course it was easier for me to extend a conciliatory hand than it was for John or Andy; in a way, Moffat's antiestablishment slurs were directed at them. Furthermore, Andy had been burnt before on matters of priority, so his tone toward Moffat was different from mine:

Thanks for drawing our attention to your earlier papers on VSL. As João already communicated to you last week, we will gladly add a comment and citation about your work. The fact that we missed your work initially is an oversight for which I apologize. I am very surprised to hear that, without even responding to João's e-mail, you approached *PRD* to raise issues of copyright. Anyone who looks at the papers will see that they are very different. I feel that we are proposing to respond to this problem in an entirely responsible way, and if you thought otherwise your first step should have been to reply to João's invitation to discuss this further with us.

Best Wishes, Andreas Albrecht

PS: Also, I do not feel you should view the publication of our work in *PRD* with chagrin. The referees were also not initially enthusiastic about our paper. We worked long and hard to build the case for its publication. Now it will give publicity to your important contributions as well.

Our "note added in proof," drafted by Andy, was also somewhat icy.

Eventually things did patch up, and I became friends with John when I visited Toronto a few weeks later. We never directly worked together, but his influence on me was immense. Ironically, he taught me how to be more conservative—the radical teaching me to be less radical! He convinced me of the importance of offending Einstein as little as possible, and I liked that. Certainly, such "minimally offensive" VSL theories were more amenable to applications outside cosmology, and I wanted to explore the more general implications of a varying c. I was starting to feel that cosmology had merely supplied the cradle for this new idea and that the time was ripe to take it farther into the world.

John Moffat's views would show me the way.

JOHN MOFFAT was born of a Danish mother and a Scottish father, was brought up in Denmark (apart from the war years), and came to be a physicist in a very unusual way. He never took a university degree and instead spent his early youth painting, for which he had a precocious talent. He lived in Paris for a while, studying with the renowned Russian painter Serge Polyakoff, and perfecting his skills in abstract art. Unfortunately, painters are often far worse off than scientists, so when he finally found himself penniless in Paris, he opted to pursue instead his other passion, physics.

Upon returning to Copenhagen, Moffat began teaching himself mathematics and physics, finding to his surprise that he had an unusual ability to absorb new concepts quickly. He made such swift progress that within a year he was working on intricate problems in general relativity and unified field theory. His work soon attracted the attention of luminaries such as Niels Bohr in Denmark, Erwin Schrödinger in Dublin, and Dennis Sciama, Fred Hoyle, and Abdus Salam in Britain, and from then on Moffat decided to devote himself fully to physics, although he never completely quit painting.

He would eventually find the right environment for his atypical background in the idiosyncratic British education system. I recall from my Cambridge days how college rules are always written in such a way that they can be broken. Everything is to be so-and-so, always "by custom and tradition," and always "at fellows' discretion," that is, it can all be overturned if a fellow thinks better of it and over several glasses of port the other fellows concur. In this vein, Sciama arranged for Moffat to matriculate for a Ph.D. without having earned an undergraduate degree. Hoyle and Salam agreed to be his supervisors, and within a year he was happily publishing papers on differential geometry and relativity. In 1958, he was awarded a Ph.D. in physics, thus becoming the only Trinity College student to matriculate without a first degree and successfully complete a Ph.D.

He then became Salam's first postdoc at Imperial College (where Salam remained for most of his life), the same place where VSL came together nearly forty years later. Moffat eventually emigrated to Canada, where he has been a physics professor at the University of Toronto ever since. When I first met him, in November 1998, he sported a clear mid-Atlantic accent and seemed fully adapted to life in Canada. He owned a remote island in Lovesick Lake, where he and his wife lived in perfect isolation for most of the year. However, his Scottish ancestry was still obvious, particularly in his facial manerisms—the way he shook his lower jaw in negatives, the deadpan blue-grey eyes, the low voice colored by long-suffering overtones.

Contrary to the popular aura he has acquired, I was surprised to find that John is actually a very conservative physicist. True, he has devoted most of his life to "alternative" theories. But his main contribution to physics is a theory of gravity that is nothing but a modernization of Einstein's last attempt at unifying all the forces of nature. He picked up where Einstein had left off, the problem being that nowadays Einstein's own approach is perceived as off the beaten track. When I first talked to John that November, I was taken aback to learn that he considered himself "the only one who actually feels that Einstein got it right." It was this belief, which I stress could not be more conservative, that had won him his reputation. If Einstein were alive now, no doubt he would be labeled the craziest of all cranks.

A few years later, Moffat would tell me that Einstein had been the first to recognize his talents, back in those self-taught days in Copenhagen. As he struggled, made progress, and eventually developed his own views on the unified theory, he had corresponded with

Einstein, who was sufficiently impressed with the young physicist to put his weight behind the launch of his career. I found it touching that Moffat's penchant for physics derived from such a beautiful personal story.

We went for a few beers, and discussed quite a lot of physics in his office on the eleventh floor of the physics tower in Toronto. Next to the pictures of Newton and Einstein that graced its walls I found one of Moffat illustrating a newspaper article titled "Challenging Einstein." That article could not have missed the point more: "In Einstein's Footsteps" is more like it.

In line with this philosophy, VSL for John Moffat was as sober an exercise as it could possibly be. He did his best to avoid conflict with relativity and its central concept, Lorentz invariance. His 1992 approach was very ingenious indeed—but I'll leave it out of this book. In 1998, when we met, John was once more active in the field and about to produce a cleaner and simpler version of his VSL theory. His leading principle was the preservation of the pillars of Einstein's relativity theory: the postulate of the relative nature of motion and the constancy of the speed of light. But how could one have a varying speed of light and not conflict with the second of these principles? It appeared to be a hopeless contradiction in terms.

John's clever approach went right to the heart of the matter and asked what the constancy of c really means. As I said before, it means that the speed of light is the same regardless of its color, the speed of its source or the observer, and when or where it was emitted or observed. But what does "light" mean in this statement? In Einstein's initial formulation it means nothing but the usual stuff you call light, not just visible light but also any other form of electromagnetic radiation, such as radio waves, microwaves, or infrared radiation. All these are exactly the same stuff as visible light, but with a frequency or color beyond the range we call "visible" because our eyes are sensitive only to this narrow band.

Light consists of particles called photons, which obviously enough move at the speed of light. According to relativity's second postulate, this speed is the same for all observers, so that mad cows running after a photon still see it moving at the speed of light. Likewise, there is no way that you can decelerate a photon until it is at rest. A box packed with photons does not make any sense. Photons exist only because they move. In some sense they are pure motion, unable to be at rest. For this reason, we say that photons have zero rest energy or mass: They are *massless*.

But herein lies the subtlety. When in relativity you talk about the speed of light, you are really talking about the speed of any massless particle, not just the photon. When Einstein first proposed his special theory of relativity, photons were the only known massless particles, but since then others have been discovered. Neutrinos are one example.* Another example is gravity itself, as Einstein would discover a few years later. The particles responsible for gravity are called gravitons, and according to general relativity it is possible to generate "gravitational light" of different colors, corresponding to gravitons with different frequencies or energies. The graviton is a particle of gravity in the same way that a photon is a particle of light. The second postulate of special relativity seems to imply that the graviton and the photon travel at the same (constant) speed: c.

Moffat's amazing realization was that the last statement is stronger than it needs to be, and is essentially unnecessary to ensure that the principles of special relativity are obeyed. It's actually possible to preserve the principles of Lorentz invariance, and therefore of special relativity, even if the speeds of the various massless particles are different. Each type of massless particle would then have its own realization of special relativity, but with different "speeds of light" for each sector. To be minimalistic (and again "conservative"), Moffat

---

*At the time of writing there is some controversy over this matter, some claiming evidence of a neutrino rest mass greater than zero.

divided massless particles into two groups: matter and gravity. The distinction arises from the general theory of relativity itself, which regards gravity uniquely as geometry. The graviton is a particle of curvature, and because it affects the structure of space-time it makes sense to set it apart from other massless particles.

Moffat then proposed that the speed of the graviton and the speed of light (and any other massless matter particles) were different. The ratio between the two was controlled by a field with a dynamics of its own, which would evolve as the universe expanded. Hence we would have a time-varying speed of light over cosmological times if we compared it to the speed of the graviton. In this fascinating fashion Moffat managed to realize VSL without insult or injury to Einstein.*

It was clever, and also very revealing of John Moffat's character. Andy and I, by way of contrast, had been totally reckless with relativity—what the hell, it's only Einstein going out the window. . . . But I was very impressed by Moffat's approach, and a few months later I would seek my own version of a Lorentz invariant VSL theory.

Even so, from my first conversations with John I got the impression that VSL had been but a detour from his main interest—his version of Einstein's grand unified theory. He felt that VSL could not be the "real thing" and that, albeit better than inflation, it was just a way of patching things up within Big Bang cosmology. If he despised inflation, he nonetheless felt that VSL was also of no fundamental importance, and I recall his referring to VSL as snake oil. Later on, he did change his views, yet this early opinion gave me an insight as to why he had given up on the publication battle, whereas Andy and I had persevered. But this is perhaps a bit unfair; Andy and I had had

---

*Moffat developed this idea in collaboration with Michael Clayton; a similar theory was proposed independently by Ian Drummond at Cambridge University.

each other, whereas Moffat had been alone. I'm sure that made considerable difference.

Another factor may well have been that John had rather complicated relations with certain scientific journals.* I quote from an illuminating e-mail I received from John Barrow in early November 1998: "João, I asked Janna Levin about Moffat as I remembered she spent time in Toronto. . . . She said that while he is very nice personally he seems to be perpetually engaged in major disputes with journals and their editors. She thought he [was] actually banned from having papers considered by some journals. . . . Sounds ideal! [The editor of *PRD*] will enjoy dealing with it."

It must be said that in his hatred for journals and their idiosyncrasies, John Moffat is in good company. Indeed, many famous scientists fell out with one journal or another at some point. Einstein is perhaps an unexpected example, but consider this incident: In the late 1930s, Einstein and Rosen wrote a groundbreaking paper on gravitational waves and submitted it to *Physical Review*. Back came a four-page report rejecting their paper. According to Rosen, Einstein was so furious that he shredded the report into little pieces, threw the debris into a bin, and kicked the bin in a rage, shouting and swearing for the next half hour. He also swore that he would never again submit a paper to *Physical Review*, and apparently stuck to his word.[†]

Talking to John Moffat over beers, I came to share his views on scientific journals. A few years later, I would write a damning article titled "The Death of Scientific Journals," curiously enough as an invited contribution to a major publishers' conference. In this article I began by describing how fraudulent scientific publishing had become. Referee reports are often empty of scientific content and reflect nothing but the authors' social standing, or their good or bad

---

*A notable exception was the "International Journal of Modern Physics."

[†]I was told this story by John Moffat, who heard it from Rosen. The funny thing is that, according to Rosen, the referee had a valid point.

relations with the referee. The senior scientists adorning the authors' list more often than not contributed nothing to the paper other than their illustrious identities—a process that smooths the refereeing process enormously. To cap it all, editors can be totally illiterate (to give due credit to the editor of *PRD*, I should add that Andy and I were truly lucky in this respect).

I then explained why despite all this corruption people still bothered submitting their papers to scientific journals: They have no choice. The establishment is set up so that one's official scientific record takes into account only publications in refereed journals, a totally artificial imposition. As a result, I myself publish all my papers in refereed journals, but regard the process cynically, as a chore not dissimilar to flushing the toilet or emptying the garbage bin. But it is also an unstable enterprise, and it contains the seeds of its own destruction. The most vocal antijournal young people I know are now becoming quite senior themselves and have not changed their views on the matter. For this reason alone, things do not bode well for scientific publishing.

But more importantly, the Web has changed everything because it has created a situation in which journals may be bypassed altogether. I have mentioned several times how physicists have started putting their papers on Web archives at the same time they submit them to journals. This has created a situation in which no one reads journals anymore because archives have replaced them. In 1992, I may have missed Moffat's paper on the Web, but nowadays, every morning before I start working I read the new arrivals in the archives. When I need a reference, I find the relevant article in the archives, display it on a computer screen, and read it there and then. It's been ages since I looked in a journal, let alone went to the library to consult one. Journals are an anachronism, and they have already effectively been phased out.

Some people think this is bad. They claim that Web archives have no quality control. That is true, but I would argue that the refereeing process associated with current journals provides no real quality control either. And in any case we don't need it; anyone should know which papers are worth reading without the need for prior filtering. Another argument is that Web archives will destroy our beloved concept of copyright. Again that may be true, but isn't the senior author in every paper already an insult to copyright? And on the few occasions someone has tried to plagiarize by using the archives it has ended badly, the person in question becoming the laughingstock of the community.

In my article I went on to argue, controversially, that this may well spread into all forms of publishing. Perhaps one day all books will be Web-based, becoming organic and ever-evolving, more copy-able and part of an environment shared by all. This may sound utopian, and in its details it certainly is, but whatever happens in the future I don't believe the printed word in the form we know it will survive the computer revolution. We should come to grips with the reality that, one way or another, the Gutenberg galaxy is dead.

OVER THE NEXT two years I kept working on VSL, if not full time then at least for a third of it. I find that it's only fun to pursue radical ideas if you mellow your research life with a few more "normal" ones. No matter what you work on, at times you inevitably get stuck, and somehow resonating between "Fringe" and "Broadway" physics is the perfect stimulant to clear your mind. And so I became a Jekyll-and-Hyde character, certainly showing only the Jekyll face when working with my graduate students: It's one thing to risk your own career with a mad idea, but quite another to ruin someone else's. Naturally over beers the Hyde character would often resurface, for their amusement—or otherwise.

Speaking of careers, in 1999 I inherited Andy's job at IC. It was very hard for me to give up the freedom associated with my Royal Society fellowship, but the fact remains that "tenure" is still the crucial turning point in science when ultimate security becomes assured. Of course this meant I had to teach, but that's been a lot of fun too.* Or rather, the peculiar zoo making up IC's physics students has made it very enjoyable. In all my time at IC I have disliked only one student, and he turned out to be a disillusioned Cambridge reject. If only IC's leaders could be more like its students.

Teaching didn't take the edge off my research, and in the two years that followed, VSL flourished. This work was done either on my own or in partnership with John Barrow. Andy dropped out at this stage, for no other reason than that he felt like doing other stuff. But Andy's departure only served to strengthen my relationship with John. Unlike most senior people, John does the hard calculations side-by-side with his collaborators, whether they be students or peers. He is also very fast, which is all the more remarkable given that he maintains a busy schedule popularizing science, giving talks in schools, writing a book every year . . . where does he find the time?

I was so impressed by his prolific output that when the Royal Society sent me a nomination form for the Faraday award (for the best "outreach," as they call it in Britain), I entered him for the competition. I praised John's many achievements in popularizing science, but was clear about the main reason I thought he was better than most other science writers: He still *did* the science! Thus in my nomination I described in great detail how he wasn't afraid to get his hands dirty alongside his younger collaborators. I argued that this was the hallmark of a true scientist and thus provided a compelling further reason to reward his popularization work.

After entering him two years in a row, and being disappointed when John didn't receive the award, I finally realized why. How tact-

---

*At least if you exclude lecturing in a full hall to one hundred students; battery chickens spring to mind.

less of me to mention John's lack of scientific impotence! My comments must have offended the entire award panel.*

Overall, these were very happy years, and perhaps the most productive period of my life. But they were tarnished by one dark cloud. In the summer of 1999, Kim decided to give up on science in an episode that distressed me a great deal. She had at the time a temporary research position in Durham with one more year to go, but things were going so badly that she opted to resign and take up a job as a high school teacher in London.

There are times in research when nothing works, and you just have to snap out of these periods—completely change your area of research, find new collaborators, new projects, and so forth. Like a snake, you have to shed your skin if you are to survive. Kim was going through one of these dramatic phases, and in normal circumstances this would simply have led to a radical change in her topic of research. However, the story turned out differently because the senior people in Durham, who should have been looking after her, decided to veto the change she wished to make.

This incident would shape my views on the nature of science forever. It made me at last conclude that physics is not like football, where it is appropriate to have two distinct types of people, managers and players. In science, managers have to be good players too, or they'll feel threatened by talent and will go out of their way to suppress it. This was plainly what happened to Kim during that depressing summer. It was also part of a pattern. A few months before, a gifted young Ph.D. student had left for similar reasons. They both fell out of science because they were miles ahead of a certain senior scientist and his idiotic assistant, and this caused resentment.

Being a woman didn't help. In Kim's own words:

---

*The first time I had to write a reference letter for one of my Ph.D. students I showed a draft to Andy, who burst into my office shouting: "Fuck, João, you can't insult the establishment in a reference letter!"

They were not opposed to the change in topic as such, but to the fact that my proposed change involved spending time in London. They argued that being close to you was my real motivation, and the science just an excuse. This is where they were most sexist in my view. I shared an office with a male researcher in the same position who was granted permission to be away from Durham for extended periods. The same people, in his case, felt that *of course* he only traveled for science—how nice that his girlfriend just happened to live in the city to which he kept traveling.

Britain has an obsession for political correctness—in what is deemed acceptable language, in the division between "good" and "bad" jokes, in people's manners and behavior, indeed, in anything nonessential and suitably superficial. In those respects I am the ultimate male chauvinistic pig and make no apologies for it. I note only that PC language and mannerisms have allowed deeply prejudiced people (as in xenophobic and racist as well as sexist) to pass as the ultimate "pro–women in science" champions. All they have to do is include the right "he or she" in their speech. And behind the scenes, where the real decisions are made, they go on being the insecure misogynists they've always been.

In this matter, as in many others, Cambridge is a great source of juicy anecdotes. I recall a meeting in Cambridge aimed at promoting women in physics, in which the men present got so excited about their own contributions to the matter in hand that they wouldn't let any of the women speak. I further recall with great fondness one of these "champions of PC-ness," who was careful to add all the "he or she" to his speech but had not yet discovered that when one stares at women's breasts, they notice it. Given his revolting looks, I was not surprised to learn that the women in question did not find this flattering. One day, I found myself in the coffee room discussing a fine point in the theory of relativity with this guy. When Kim walked past, in line with his usual behavior, his eyes followed her arse. I was already going out with Kim at this stage, so in a surge of Latinism I said: "Good, isn't it?" Needless to say, he has avoided me like the plague ever since.

Kim's leaving science was central to me during the years in which VSL came together. I became much less solicitous of the establishment, and some of the more extreme views voiced in this book formed in my mind at about this time. And VSL's developments were shaped in part by a physical need to insult the hypocrisy and corruption of the scientific establishment.

Somehow this angry energy was just what I needed. VSL really took off. And the views from above the clouds were of such redeeming colors that I have to admit, in spite of everything, that those two years were truly joyful.

OVER THIS PERIOD, my VSL work reflected foremost John Moffat's influence as I endeavored to find ways to reconcile VSL with relativity. I did this not out of fear of contradicting relativity, but because I was enticed by how much more easily these "conservative" VSL theories could be applied outside cosmology. I was ready to expand my range of interests, and it was extremely difficult to do so with the theory Andy and I had first proposed. At the time, this hadn't bothered us because we were merely trying to find a competitor to inflation, and inflation certainly had nothing to say outside cosmology. But by this stage my standards had been raised; I expected VSL to outdo inflation by predicting something about the physics of the current universe rather than being just a short episode in the life of the very early universe.

I therefore started a process, which has continued to this day, of converting VSL from a single theory into a large class of models. Until experiment proves just one of these models true, we should play with them all. Likewise, there are hundreds of inflationary models, and such a state of affairs will remain in force until one of them is conclusively verified.

I eventually produced my own version of a Lorentz invariant VSL theory, which was far from easy. But my efforts paid off and the new theory led to an amazing array of predictions.

As Moffat had done before me, I carefully scrutinized the small print in Einstein's second postulate, searching for other avenues towards a Lorentz invariant VSL theory. While doing so, I recalled a discussion Andy and I had had with the *PRD* editor, who had cleverly questioned whether a varying c could ever be an observable phenomenon. He had noted that by changing, say, the way time is measured (that is, the "units" of time), we could always enforce any variation (or lack of it) in the speed of light. But if a result depends on the choice of units, then clearly it cannot represent an intrinsic aspect of reality.

The editor had used this argument to attack the idea of a varying c, but I later realized that the logic could be reversed to attack the constancy of c along exactly the same lines. Seen from this perspective, it would seem that postulating the constancy of the speed of light is nothing but a convention, a definition of the unit of time, which, in turn, ensures that the postulate is true. Is Einstein's famous postulate a tautology?

The answer is both yes and no. I quickly realized that there *are* aspects of the second postulate that depend on the units you choose, but there are also others that do not. Einstein's cows did perform a real experiment (or rather, Michelson and Morley did), so the second postulate cannot be totally vacuous. And indeed, when for instance I state that the speed of light does not depend on its color, this doesn't depend on the units I use. If I take two light rays with different colors and measure their speeds at the same place and time with the same clocks and rods, when I take their ratio I will always find that the result is one, regardless of what units I have used. The ratio of two speeds, like pi (which is the ratio of two lengths), has no units and is therefore always the same regardless of what clocks and rods are used. Therefore, when confronted by the destructive argument raised by the *PRD* editor, this particular aspect of the second postulate of relativity is unassailable.*

Other aspects of the postulate, however, are not impregnable and are indeed tautologies or conventions. Very relevantly, stating that the speed of light at different times and places is the same *must* depend on how I decide to build my clocks. And how can I be sure that their ticks are the same everywhere and at all times? Such a "fact" has to be a definition, a tacit agreement made between all physicists. To be more concrete, in the context of changing alpha theories, electronic clocks are just like pendulum clocks and tick imperceptibly "differently" on Earth and on the Moon. Hence by stating that the speed of light is the same at all times and everywhere, we are effectively making the same mistake as taking a grandfather clock on a spaceship.

Therefore I recognized that while part of Einstein's second postulate is physically meaningful, other parts are not and cannot reflect the outcome of any experiment. I decided to keep the essential stuff and throw away the rest, which left enough room for the possibility that the speed of light varies in space and time. The result was a Lorentz invariant VSL theory. In this theory, at any given point in space-time, the speed of light does not depend on its color or direction, or on the speeds of either the emitter or the observer. The outcome of a Michelson-Morley experiment is still the same as it is in special relativity, and the value of c at a given point still represents the *local* speed limit. However, the value of this speed limit may vary from place to place and from time to time. These features are certainly not valid for all flavors of VSL theory, but I decided to stick with the "vanilla" version for a while.

---

*Notice that Moffat's VSL theory is also immune to this attack. It states that the ratio of the speeds of photons and gravitons varies in space and time. This ratio has no units and is therefore independent of which clocks and rods I decide to use.

Those two blissfully uncomplicated years playing with this "conservative" VSL theory were essential for my confidence. I was finally able to formulate VSL using a principle of minimal action: Maupertuis was back in business. But more importantly, this new version of VSL could more easily be applied to branches of physics other than cosmology. This led to an explosion of new predictions and interesting features, and these further stimulated my enthusiasm for my pet theory.*

For example, when I investigated the physics of black holes under VSL, I found several surprising results. Black holes are a puzzling prediction of general relativity, objects so massive and compact that light, or indeed anything else, cannot escape them. According to relativity, like other objects, light "falls" toward massive objects in its vicinity. Rockets with their engines switched off fall towards the Earth—and so does light. But rockets have an "escape velocity," a speed above which they may leave the Earth's influence, but below which they will forever be confined by its attraction. The escape velocity for a black hole is greater than the speed of light!

But let me be more precise. The escape velocity depends on two things: how massive the attracting object is, and how high you are. The first is obvious; you'd need a larger thrust to leave Jupiter than Earth. But also if a rocket is already orbiting the Earth, it needs a smaller thrust to get away than if it is still on its surface. The more rigorous definition of a black hole is an object for which there is a "height" (or distance to its center) below which the escape speed becomes larger than c. Since nothing can move faster than light, if you find yourself below this height, then you're stuck for good.

Thus black holes must be both massive and compact, so that this point of no return lies outside their surface rather than being buried inside them. The region where the escape speed becomes the speed

---

*⁰I should stress that many of these discoveries are peculiar to Lorentz invariant VSL theories, and are not valid for other approaches.

of light is called the "horizon" of the black hole. As with its cosmological cousin, the black hole horizon represents a curtain of ignorance. For those outside it, the horizon defines a surface beyond
which lies the unknown, since nothing can cross it from inside to tell
us what is going on. The interior of a black hole horizon is permanently disconnected from us.

A black hole is "black" because, even if the stuff inside it shone,
its light would fall back into the black hole like fireworks fall back to
Earth. For this reason, we can never hope to see a black hole directly.
All we could ever see would be spaceships about to cross the horizon, braking like mad, their doomed crews radioing frantic SOS signals, then, all of a sudden ... perfect silence. Not because their
equipment broke down, but because their cries for help are now
being gulped along with them, both falling uncontrollably toward the
voracious black hole.

And what could be the role of a varying c in all this? I soon found
that in VSL theories the speed of light varies not only in time, as the
universe evolves, but also in space. Near planets and stars the effect
is almost imperceptible, but close to a black hole something more
drastic could happen. To my great horror, the equations led inex-
orably to the conclusion that, at the horizon, the speed of light might
itself become zero!

This result has tremendous implications. It shows that some VSL
theories predict that you may be prevented from entering a black hole
horizon. According to conservative VSL, just as in special relativity,
the speed of light is still the speed limit—it's just that this can vary
from road to road. Your speed must always be smaller than the local
value of c, so if the speed limit falls to zero you have just run up
against the ultimate red traffic light. You have to stop at the horizon
of a VSL black hole. At the very edge of the precipice, your suicide
attempt is thwarted. VSL black holes are sealed off against disaster.

Another way of explaining this curiosity is to note that great irregu-
larities occur in electronic clocks near VSL black holes. Whatever

way we decide to define time, these clocks would tick differently near a black hole. But biological processes are themselves of an electromagnetic nature, which means that the rate at which we age is in fact an excellent electronic clock. I found that near a VSL black hole, we would age much faster, not because of Einstein's time dilation effect, but because the speed at which electromagnetic interactions take place becomes higher. Therefore, as we approached a VSL black hole, our heartbeats would speed up and we would age more rapidly; or, conversely, we would see our motion towards the horizon slow down, as measured by our pace of life. As we closed in, an eternity (as measured by our inbuilt clocks) would go by while barely one second would have passed if c had remained constant. The horizon would be closer, but also more unreachable. The horizon of a VSL black hole is like a goal at an infinite distance, an inaccessible edge of space, beyond which lies a curious box of eternity.

I found this exotic enough, but this "conservative" VSL theory would have even more startling implications. Once I realized that c could vary in space as well as in time, I set off to study what other kinds of spatial variations were possible. One in particular was mind blowing: "fast-tracks." These are objects that occur in some VSL field theories, taking the form of cosmic strings along which the speed of light is much higher.

Cosmic strings are hypothetical objects predicted by some particle physics theories. In fact they are not dissimilar in origin to the magnetic monopoles that so worried Alan Guth. But whereas monopoles are point-like, cosmic strings are line-like: They are long threads of concentrated energy extending across the universe. To this day, cosmic strings—like black holes or monopoles—have yet to be observed, but they are a logical prediction of very successful particle physics theories.

When I plugged cosmic strings into the equations of this VSL theory, a complete monster emerged. I found that the speed of light could become much larger in the immediate vicinity of the string, as if a "coating" of high light-speed enveloped it.

This would create a corridor with an extremely high speed limit extending across the universe. And this is just what space travel is begging for: a fast lane. But it's even better than that! Recall my earlier mad cows and how they stayed young as they moved about at vertiginous speeds, while the sensible farmer got older every day. Einstein's time dilation effect creates a terrible predicament for space travel. Even if we found a way to travel close to the speed of light, although it might become possible to make a return trip to distant stars within a single lifetime, when the spaceship returned, its occupants would find their civilization gone. For while a few years had elapsed for the travelers, millenia would have flown by on Earth.

Along a VSL cosmic string no such annoyances would hinder the space traveler. Of course, in these VSL theories there is still a time dilation effect, since the theory still satisfies Lorentz invariance. But, as with special relativity, this effect only becomes significant if the speed of the traveler is comparable to that of light, which in this theory means the *local* value of c. Since along a VSL cosmic string the value of c may be much higher, we could move at very high speeds indeed and still be traveling much more slowly than the local value of c, so that time dilation would be negligible. The enterprising astronaut could then move speedily along fast-tracks, exploring the most distant corners of the universe but still moving much more slowly than the local speed of light. He would thus avoid the "twin paradox" effect, and on his return he would still be roughly the same age as his twin. Not only could he visit distant galaxies within his own lifetime, but he could return home within the lifetimes of his contemporaries.

This is undoubtedly a bewitching implication of VSL, and if true it will dramatically change the way we perceive ourselves in the universe, as well as our prospects of contact with alien life. But perhaps the most striking development concerns the overall picture of the universe associated with these theories.

Einstein originally introduced the cosmological constant into his theory to make the universe static and eternal. Like many scientists then and now, Einstein was deeply troubled by the idea of a universe with a definite beginning (even if it was several billion years ago). After all, what happened before the Big Bang: In effect, what banged? Is it meaningful to talk about the "beginning" of time itself? For Einstein, and for many others after him, an eternal universe made much more sense philosophically.

But the static universe simply could not withstand Hubble's observations, and Einstein would later repudiate the tool he had employed to achieve his goals: the cosmological constant. For the next several decades, wishful thinking alone kept Lambda away from most cosmological considerations. Little did Einstein and his peers know in just what twisted way Lambda would reemerge on the stage of cosmology at the end of the twentieth century.

One such instance was inflation, but yet another surprise was in store. Since Hubble's discovery of the cosmic expansion, similar astronomical observations have been made with ever-improving accuracy. In particular, for the past few years, astronomers have been studying supernovae in far distant galaxies in the hope of discovering how fast the universe was expanding in the distant past. The aim is to throw light on the question of how rapidly the universe is slowing down, as it should if gravity is as attractive as it seems to be.

But the result seems wholly paradoxical: It appears that the universe is expanding at a faster rate today than it did in the past—the cosmic expansion is *accelerating*! It could do this only if some mysterious repulsive force is pushing the galaxies apart against the natural tendency of gravity to pull them together. Of course, theorists are familiar with the idea of such a force. It is Einstein's cosmological constant, Lambda, rearing its ugly head again.

This is an unexpected twist. It seems that the cosmological constant is not zero. But if the vacuum energy is a significant component of our universe after all, why is the universe only recently beginning to feel its effects? As we have seen, Lambda likes to dominate, so if it exists at all it should have overwhelmed ordinary matter long ago, blasting all the galaxies to infinity. So why is the universe still here?

One possible solution is VSL. We have seen how a sharp decrease in c converts the vacuum energy into ordinary matter, solving the cosmological constant problem. It is now possible to make the dragon bite its tail and construct a dynamical theory in which the cosmological constant itself is responsible for changes in the speed of light. In this perspective, every time the speed of light decreases sharply, Lambda is converted into matter and a Big Bang occurs. As soon as Lambda becomes nondominant, the speed of light stabilizes and the universe proceeds as usual. However, a small residual Lambda remains in the background, and eventually it resurfaces. According to VSL, astronomers have just observed the reemergence of the cosmological constant.

But as soon as this happens, Lambda proceeds to dominate the universe, thus creating conditions for another sharp decrease in the speed of light—and a new Big Bang! The process goes on forever, in an eternal sequence of Big Bangs.

Strangely, and rather beautifully, it is possible that the varying speed of light theory yields an eternal universe with no beginning and no end. The future for the universe we see now is rather bleak. As the Lambda force grows, it will push all the matter in the universe away to infinity. The sky will darken as the galaxies disperse and become lonely beasts lulled into oblivion in a sea of nothing. But under these barren conditions, VSL predicts the generation of vast quantities of energy out of the vacuum. Thus the empty, vacuum-dominated universe provides the conditions for a new Big Bang, and the cycle begins all over again.

Paradoxically, if the speed of light does change, then the universe itself is eternal, and Einstein's greatest blunder will turn out to be his greatest vindication.

BUT THESE EXHILARATING discoveries were not the end of it. Once I realized there were many possible VSL theories, and that these theories have implications in all areas of physics, I was ready to become radical again and contemplate the full effects of breaking Lorentz invariance. My new confidence arose when I became aware that VSL had something to say about *the* ultimate physics conundrum, the sort of puzzle string theory attempts to address. And I was ready for a walk in the wild.

12   ALTITUDE SICKNESS

YOU MAY BE SURPRISED to learn that Einstein died deeply frustrated with his achievements. It's easy to dismiss this grievance as megalomaniacal high standards, but there was also some basis for it. Throughout his life Einstein had pursued mathematical beauty, conceptual simplicity, and, above all, cosmic unity. Just think of the brainwaves behind the discovery of the oneness of mass and energy, or the remarkable explanation he gave for the equality of inertial and gravitational mass. Indeed, all his theories seem to revolve around the search for unification: bringing concepts together under one larger, better-designed, and prettier umbrella.

But then, in his early forties, he got stuck in what would become a lifelong obsession. He had gotten stuck before, but this time he would die without solving his riddle. The unsolved mystery that finally beat Einstein was the quest for the grand unified theory of electromagnetism and gravity—the *theory of everything,* as we now like to call it. But this final search for unified beauty led only to a terrible mess, as more and more types of force were found (such as the weak and strong interactions, mediating nuclear reactions), and unwanted technical complications started to pile up.

Just to make matters worse, the problem gradually transmuted into the need to unify gravity and quantum mechanics. We know that we live in a quantum world: Energy can exist only in multiples of elementary units, called quanta; uncertainty plagues theories and observations whenever they attempt to examine extremely small amounts of matter containing only a few quanta. Electricity and its twin, magnetism, are known to be quantized, and the photon turns out to be the finite particle that represents the elementary units of electromagnetism. The weak and strong forces are also quantized—that much is well understood.

By contrast, no one has ever produced a proper theory of "quantum gravity," and the graviton—the quantum of gravity—remains ill-understood and elusive; thus unifying gravity with the other forces of nature seems futile at this stage because we can't have a unified theory in which one half is quantized and the other half isn't.

Quantum gravity has become a massive brainteaser, a bit like Fermat's last theorem or other nightmare problems scientists have given themselves. Will this be VSL's final showdown?

As usual, to fully understand the problem requires comprehending a number of technical matters that are clear only to experts. But the crux of the puzzle is easy to explain in plain language. Since that decade of hard labor that led Einstein to general relativity, we have known that gravity is a manifestation of the curvature of space-time. Space-time is no longer a fixed background where things happen; rather, it can bend and warp, so that its landscape evolves in intricate patterns, which themselves are the dynamics of gravity.

Quantizing gravity therefore means quantizing space and time. There should be indivisible, smallest amounts of length and duration: fixed quanta making up any period or separation. Such quanta are called the Planck length (L_p) and Planck time (t_p), and no one really knows what they are except that they must be minute.

But even before we think too much about it (and we shall return to this matter very soon), it should be obvious that to quantize space and time we need an absolute clock and rod, two concepts that special relativity has denied us. If indeed space and time are to be granular, then their atoms ought to be absolute; but there is no absolute space or time. We're left handcuffed by our own devices. We have quantum theory on the one hand, and special and general relativity on the other, and we are asked to produce a theory of quantum gravity using their precepts. But what comes out is a contradiction.

I should stress that the need for a theory of quantum gravity does not arise from a conflict with experiment because we have yet to find a physical effect ruled by quantum gravity. It could be that there *is* no unification and that gravity simply isn't quantized. But this possibility seems to insult our sense of logic. Nature is crying out for a single principle capable of containing within itself the current shambolic variety of theories we use to explain the physical world around us.

Furthermore, we have encountered the quantum gravity mystery before, in the definition of the so-called Planck epoch, that period in the life of the hot young universe when it expanded too fast to be properly understood without invoking quantum gravity. Thus, in this sense, the search for quantum gravity is the search for our origins, hidden deep inside the Planck epoch. But now we learn that this box of ignorance—the Planck epoch—is part of a larger conundrum. It's part of Einstein's deathbed vexation, his unfinished symphony. Einstein's last words, uttered in German, were lost on his American nurse, but it's very possible they were something along the lines of: "I knew that motherfucker would beat me in the end."

TODAY WE ARE none the wiser than Einstein as he let out his last sigh and said whatever he had to say. Nearly fifty years on, physicists look with dissimulated disdain at Einstein's last efforts (the so-called nonsymmetric metric theory of gravity), as if they belonged to a senile old man. But no one likes to admit that our own meager efforts are contemptible, to say the least. I like to picture God wetting his (or her!) pants in hysterical laughter while contemplating all the crap we have come up with as theories of quantum gravity.

But what we lack in achievement we make up for in panache. Indeed, we now have not just one "final answer," but at least two, and although no one has the faintest idea how to test these theories with current technology, everyone is quick to claim that he alone holds the holy grail and that the others are all charlatans.

The two leading quantum gravity cults are called string theory and loop quantum gravity. Since they don't connect with experiment or observations at all, they have become fashion accessories at best, at worst a source of feudal warfare. Today, they constitute two enemy families, and if you work on loop quantum gravity and go to a string conference, the local tribe will look at you in amazement and ask what the hell you're doing there. Assuming you are not boiled in a cauldron, you then return home, only to be scolded by your horrified loop colleagues, who think you are out of your mind.

As with every cult, people who do not conform to the party line are ostracized and persecuted. When a bright young string theorist wrote a paper providing dangerous ammunition to the loop people, a stringy little bitch commented that "if he writes another one like that he'll lose his string theory union card." A mob mentality has developed, and being tagged "string" or "loop" opens or closes doors accordingly. If you acquire the loop label, don't even think of applying for a string job.

Great animosity, even visceral hatred, has developed between these factions. The "final answers" provided by religious fanatics, with their singular panoply of denominations, come to mind. The world would be so much better without religious fundamentalism, whether it be of the scientific or nonscientific variety. Sometimes I think the existence of these people provides the best proof that God does not exist.

Unfortunately, Einstein himself bears a lot of the responsibility for having brought about this state of affairs in fundamental physics. The young Einstein set himself the task of eliminating from his theories anything that couldn't be tested by experiment. This commendable view turned him into a scientific anarchist, demolishing absolute space and time, the ether, and many other fantasies clogging up the physics of his day.

But as he grew older, he changed his mind. He became more mystical and started to believe that mathematical beauty alone, rather than experimentation, could point scientists in the right direction. Regretfully, when he discovered general relativity—employing this strategy—he succeeded! And this experience spoiled him for the rest of his life: It broke the magic link between his brain and the universe, the link that would seek insight from experiment only. And indeed, after general relativity he produced little of value, and became more and more detached from reality.

Today, it is the old Einstein whom scientists working on quantum gravity try to emulate in their sterile belief that God-given beauty, rather than experiment, will point them in the right direction. This obsession with formalism is, to my mind, what has derailed generations of scientists working on this topic. In a sense, they seem to adore the old Einstein, not noticing that the younger one would certainly despise his older version, and that maybe it's the younger guy we need to follow—if we follow anyone at all.

When John Moffat first visited Niels Bohr in the 1950s, after corresponding with Einstein on his grand unified theory, Bohr told him: "Einstein has become an alchemist."

IT IS PERHAPS not surprising that VSL has something to say about quantum gravity. After all, it shakes the foundations of physics, and the puzzle of quantum gravity is as fundamental a physics matter as one can find. The story is very different with inflation, which has nothing to say about quantum gravity. If anything, inflation proponents have tried, and failed, to derive it as a side effect of quantum gravity. Their hope is that inflation might naturally arise during the quantum epoch, but no one knows how this could come about. In contrast, VSL unavoidably changes the perspective of quantizing gravity. This realization led me to explore further versions of the theory, with direct implications for quantum gravity models and string theory.

I don't wish to waste too much time on the details of current theories of quantum gravity, but let me give you just a taste of their vagaries. One major attempt at quantum gravity and unification is string theory, which in recent years has been revived in the form of something called M-theory. According to followers of the M-persuasion, the world is made of strings rather than particles (with, in recent years, strings being replaced by membranes and other objects, too). The length of such strings is usually identified with the Planck length mentioned above, so that for most practical purposes strings are indistinguishable from particles.

At a fundamental level, however, a stringy world is very different from a particle world, and there are two main reasons why strings may be seen as desirable. The first is that one may expect unavoidable space-time quantization to emerge in such a world. Indeed, if the smallest objects making up matter have a nonvanishing size, then regions smaller than them become essentially metaphysical, because we lack a thin enough scalpel to dissect them. Given this effective space-time quantization, it is not surprising that many of the technical difficulties associated with the quantization of gravity evaporate in a stringy world. And indeed, string theory is not a bad attempt at quantizing gravity.

Another good reason for strings is their ability to unify seemingly different particles and forces. The strings of a guitar vibrate in a variety of harmonics; so, too, "fundamental strings" (as they're called) can be plucked in their own musical scale. With each note, the string acquires different qualities, storing different amounts of vibrational energy. Far away from the string, observers can't distinguish the vibrating object but only see something that looks like a particle. The stunning realization made by string theorists is that for such an observer, each note corresponds to a different type of particle.

This could be the ultimate unification scheme! Photons, gravitons, electrons, and so forth—all particles and forces would be nothing but different configurations of one type of object: fundamental strings. It's a beautiful vision, like many other aspects of string theory.

This would all be very well, were it not for the fact that they never tell you it's all "work in progress." The truth is that they have not yet consistently quantized space-time or curvature; indeed, they are unable to see space-time in the same relative manner that Einstein saw it: Strings live on a fixed background space not dissimilar to Newton's clockwork universe. Another catastrophic failure is the stringy musical scale—their music may be the sweetest harmony of the heavens, but it has nothing to do with the real world. Its predicted lightest particle (after the photon, graviton, and other massless particles) is billions of billions of billions of times heavier than the electron. String grand unification remains wishful thinking.

But the stringy disasters don't end here. In the 1980s, string theory could work only in a world having twenty-six dimensions. Then a revolution came along and they started to work in ten, two, and (make sure you don't fall off your chair) *minus* two dimensions. Today they work in eleven dimensions. But string theorists are unabashed: Whenever someone has the cheek to propose a theory that works in three spatial dimensions and one time dimension, they dismiss it as obviously wrong.

This is bad, but to my mind perhaps worst of all is that in the fine details there are thousands of possible string and membrane theories. Even assuming that someone does eventually find a theory that explains the world as we see it, with all its particles living in four dimensions, one may then ask: Why *that* theory and not one of the others? As Andy Albrecht once exploded, string theory is not the theory of everything, it's the theory of anything.

This criticism is nowadays countered by the remark that in recent years all of these string and membrane theories have become unified in a single construction: M-theory. M-people say this with such religious fervor that it is often not noted that there is no M-theory. It's just an expression referring to a hypothetical theory no one actually knows how to set up. To add to its mystique, the cult leader who coined the term never explained what the M stood for, and M-theorists heatedly debate this important issue. M for mother?

M for membrane? M for masturbation seems so much more befitting to me.

All in all, I don't see why so many impressionable young scientists fall for the supposed charms of M-theory. Stringy people have achieved nothing with a theory that doesn't exist. They are excruciatingly pretentious in their claims for beauty; indeed, we are all assured that we live in an elegant universe, by the grace of stringy gods. Personally, I don't find aesthetic appeal enough, and I think it's about time someone points out that the king, walking down String Avenue wearing glorious M-garments, is actually naked.*

I HAVE TO ADMIT THAT, in spite of all this, I'm not immune to the mathematical beauty of string theory. Indeed in the summer of 1990, before turning my interests to cosmology, I started a Ph.D. in string theory. But I quickly became depressed with the utter lack of prospect for contact with experiment. All I could see was a mafia of self-indulgent pseudomathematicians, throwing Masonic jargon at each other in an attempt to hide their lack of scientific achievement. I gave up on string theory, became a cosmologist, and never regretted this wise career move. It is therefore ironic that ten years later I would find myself tangled in strings once again.

The string theorist who brought this about was Stephon Alexander, who became a postdoc at Imperial College in the autumn of 2000. Stephon was not like other string theorists; he had a wide-open mind, full of visionary flights, and above all an exuberant personality.

Stephon was born in Moruga, Trinidad, but his family moved to the United States when he was seven. He was mainly brought up in the Bronx, at a time many people were desperately trying to bring about change in the poorest areas. Bright kids were being put on special programs, and charismatic headmasters were moving in to face

---

*In addition, picture a world full of strings instead of particles. How can a universe pervaded by cosmic pubes be considered more beautiful?

STEPHON ALEXANDER

the challenge. Stephon benefited from much of this, and after graduating from the De Witt Clinton High School, was offered scholarships by several Ivy League universities. A talented jazz sax player, Stephon nonetheless opted for a career in physics. He earned his degree at Haverford, followed by a Ph.D. in physics at Brown, where his thesis adviser was Robert Brandenberger, a cosmologist and longtime friend of mine. But Stephon quickly moved into string theory, avidly reading its vast literature.

Still a Ph.D. student, Stephon started a new line of intriguing research showing how VSL could be associated with M-theory. Before he had time to write it up, Elias Kiritsis, working independently in Crete, published the same idea. This is a common misfortune afflicting graduate students working in busy areas of physics. But as often happens in such dramas, Stephon's work was sufficiently different (more developed in some ways, less so in others) that he was able to publish it.

Their idea was brilliantly simple. As I said before, M-theory is not just about Planck-sized strings (which are linear or one-dimensional objects), but also about membranes (which are planar, two-dimensional objects). Indeed, once you realize that M-theory likes to live in eleven dimensions, it becomes clear that all manner of higher dimensional objects are allowed (which, in the M-jargon, are called p-branes).

The space-time we see, however, is four-dimensional. Since Kaluza and Klein, we know that it's possible to reconcile these two statements by curling up the extra dimensions into circles of very small radii, so that we can't perceive them. But it could also be that we live on a 3-brane, that is, a large, possibly infinite three-dimensional membrane to which a time dimension is added. This so-called "brane-world" cosmology does not require the extra dimensions to be small, merely that we are somehow "stuck" inside this 3-brane, which happily floats around in an eleven-dimensional space. Of course, we have to explain why matter can't leave the brane, or else we would leak into the extra dimensions. But a variety of mechanisms have been put forward, confining the type of matter we are made of to a 3-brane.

Kiritsis and Stephon examined life on one such a 3-brane that happened to be moving in the vicinity of a black hole. They assumed that the speed of light in the full eleven-dimensional space is a constant. But when they studied the motion of light "stuck" to the 3-brane, they found that its speed varied! In fact they calculated that the speed of light as seen on the brane was simply related to its distance to the black hole, and that as the brane approached the black hole, VSL would be realized. This avoids direct conflict with relativity, since the fundamental, eleven-dimensional speed of light remains constant. However, it produces VSL if all you know about the world is the three-dimensional membrane that you call the universe.

For me, these papers were a blast from the past in that they were very similar to where I started, in January 1997, with a pub joke while chatting to Kim. I had played with Kaluza-Klein VSL before moving on to other ideas, and now string theorists were playing with precisely the same type of theory. Although I dislike the cult aspects of string theory, I'm not religious about it. I was therefore very happy to start working with Stephon on possible realizations of VSL in M-theory.

In October 2000, Stephon moved to Imperial College and we quickly became good friends. He soon found a new home in Notting Hill, blending into its large Caribbean community at once. In spite of all the recent gentrification, Notting Hill is still a great place to live. And the reason is simple.

In 1944, in a last desperate attempt to break the English morale, the German army bombed the shit out of London with the first successful missiles, the V1 and V2 "flying bombs." The effects were devastating, far beyond anything conventional bombers had thus far been able to achieve. Entire blocks were wiped out wherever a missile landed. In particular, great havoc was caused in the historical parts of London.

After the war, with the country in tatters, few people thought of rebuilding whole blocks in the nineteenth-century stucco style typical of old London. Maybe there was enough money to restore a house here and there, but where a V1 or V2 had caused large-scale damage there now sprouted concrete or redbrick monstrosities. So much so that these days a casual stroll in central London is sufficient to reveal the sites where missiles hit the target.

Ironically, little did Adolf Hitler know what a service to democracy he was providing. In the 1950s and 1960s, as the British welfare state took off, these monstrosities became council housing estates, low-rent accommodation for the disfavored. This has effectively acted as an antighetto device, ensuring that places such as Notting Hill contain quite a mixture of rich and poor. Trustafarians with unfounded arty-farty hopes nowadays rub shoulders with (mostly) impoverished Caribbean, Irish, Moroccan, and Portuguese communities.*

I introduced Stephon to the Globe, a Caribbean late-night dive, only to find out a week later that he knew more dens than I had ever dreamt of. But it would be at the Globe that Stephon and I would carry out most of our discussions on M-theory realizations of VSL. Imperial College at this stage was truly overburdening me with all manner of crap, so I got away as much as I could. In contrast, the relaxed atmosphere of the Globe proved most congenial to our wanderings. We were quite often on a high.† This was a scientific period during which my work suffered from what might be called "altitude sickness."

As the months unraveled, Stephon became known in the Globe as "The Professor." In our digressions we were often assisted by "The Eagle," a resourceful Jamaican always keen to help us. Speculating in this sort of environment and "frame of mind" is good and bad, to be fair. Sometimes it's a bit like dreaming: While you're asleep it all works out beautifully, but as soon as you awake, if you remember anything at all, you'll often recognize what a pile of rubbish you've been thinking up. Einstein's cows are a very unusual occurrence. Similarly, Stephon and I had many false starts, but this period was a lot of fun. As Stephon pointed out, one cannot rush creativity.

And one day our elevations did lead somewhere concrete. Stephon became interested in a connection between M-theory and something called noncommutative geometry. This is a version of geometry in which space-time appears "atomized." We examined the motion of "photons" in such spaces and found a striking result. For light with a wavelength much larger than the size of the granules of space, noth-

---

*Notting Hill, the famous film set in our neighborhood, required severe bleaching: A scene down Portobello Road managed not to feature a single black person. My mathematical mind actually estimated the probability for this to happen "by chance." The odds are not quite as low as the universe being flat by chance, but are nonetheless very suspicious.

†Metaphorically, of course.

ing unusual happened. However, for very large frequencies (that is, very small wavelengths), light would start to notice that it doesn't live on a continuum, but instead has to leapfrog over the potholes. This causes an increase in its speed, the more so the larger its frequency. In effect we found that the speed of light in noncommutative spaces is color-dependent and increases at very high frequencies. We had glimpsed yet another realization of VSL.

Our idea was therefore to realize VSL cosmology *indirectly*. In our model, the speed of light was not time-dependent per se, as in the model Andy and I had proposed. Instead, we used the Hot Big Bang itself to drive changes in c. Going back in time, towards the birth of the universe, the cosmic plasma would become hotter and hotter, meaning that the energies, or frequencies, of your average photon would also become higher. Eventually, the frequencies would be high enough for the phenomenon of frequency-dependence in the speed of light to kick in. Hotter plasma then translated into a larger ambient speed of light in the universe. And so we had a varying speed of light in the early universe, not because the universe was young, but because it was hot.

A very strange period followed in which Stephon and I went in opposite directions at every turn. Stephon wanted to relate our work more closely to the intricacies of M-theory; but I knew that this would lead us away from the realm of observations. So I tried to make our model more "down to earth," capable of making cosmological and observational predictions. In turn, this forced me to introduce wooden, arbitrary assumptions, which totally horrified Stephon's stringy sense of beauty. In this conflict, Stephon and I were experiencing a common tragedy—no one knows how to do quantum cosmology, that is, how to combine quantum gravity with cosmology, thereby bringing the experimental successes of the latter to bear on the former.

What came out in the end is best described as a cross between a horse and an elephant—neither fish nor fowl—a sort of mule with a trunk. Unsurprisingly, we got very mixed reactions, from string theorists and cosmologists alike. We didn't give a shit of course; for us, those crazy papers will forever be associated with the Globe ambience and the virtues of altitude sickness. But I learnt an important lesson: If you play the United Nations, you will get fired at from both sides. This is endemic to current cosmology and quantum gravity. Will they ever meet, one day?

Andrei Linde once described the prickly relation between cosmology and quantum gravity with an interesting metaphor, based on a real incident. In the days of the Soviet bloc, plans were drawn up for an underground train line linking two areas of a major capital in Eastern Europe. Work eventually started, and tunnels were dug from both sides. However, as the work progressed, it became obvious that the survey work carried out beforehand was hopelessly inaccurate and that there was no guarantee that the two tunnels would ever meet!

But resourcefulness was certainly a major asset of the Soviet era, and a renewed green light was promptly given by the planning teams. The logic behind this decision was simple. If the tunnels happened to meet, one would end up with one train line as initially planned; if they didn't, one would instead have built *two* train lines.

Regarding cosmology and quantum gravity, we all feel the same way. We are certainly making progress from either side. Yet sometimes . . . I fear the worst.

BUT IF I FLIRTED with the Montagues, I didn't overlook the Capulets, and perhaps the closest link between VSL and theories of quantum gravity resulted from my collaboration with Lee Smolin, one of the creators of loop quantum gravity.

Lee moved to Imperial in autumn 1999 as a visiting professor, bringing along a large entourage of postdocs and students. While in London, he led an independent life, doing most of his work in cafes and rarely being seen in his office (coincidentally the same little office that, a few years before, had witnessed the long nights that shaped VSL). For this reason we did not properly meet for nearly a year.

LEE SMOLIN

At first, Lee was not aware of VSL and we ended up working together only through a quite foreign influence. I should say that Stephon and I were not the first to propose a color-dependent speed of light (although we were definitely among the first to construct a cosmological model based on it). Giovanni Amelino-Camelia in Italy, Kowalski-Gilkman and others in Poland, as well as Nikos Mavromatos, Subir Sarkar, and many others in England, had all entertained the concept of an energy-dependent speed of light in many different theories of quantum gravity.

It was these people, Giovanni in particular, who planted VSL in Lee's mind. What enticed him most was the prospect of VSL effects placing quantum gravity theories in the courtroom of experiment within the next few years. Unlike most other quantum gravity scientists, Lee did not believe that God was about to give him a blow job—that his little theories would turn out to be true just because they were "elegant." Instead, Lee wanted experiment to flood quantum gravity, allowing nature to decide for itself. Therefore, rather than becoming dismissive or defensive, his eyes would gleam when someone said that quantum gravity could soon be tested. And that's how we started working together.

Our work was based on a disconcertingly simple premise. We knew that quantum gravity should predict new phenomena or observations. But for once, against all trends in the field, we were modest; we assumed that with current technology there was no way in which to probe these effects. There was only one thing we knew for certain: At the paltry low energies currently accessed by accelerators, or at the large space-time scales for which our sensors can probe curvature, *no* quantum gravity effects have been seen, and classical gravity (i.e., general relativity) is at the very least an excellent approximation to the real world.

Our only assumption, therefore, was the existence of a *threshold,* a border beyond which new effects, belonging to the ultimate theory of quantum gravity, would indeed become significant, but below which they would remain negligible. An energy, called the Planck energy (E_p), should mark this threshold, and the new effects should be present only above E_p. Similarly, there should be a length, the Planck length (L_p), which tells you how much magnification a quantum gravity microscope would need before it became capable of seeing the discrete nature of space and curvature. Finally, a duration, the Planck time (t_p) will tell you how short-lived these new effects should be.

In fact Lee and I didn't even need to know the values of E_p, L_p, or t_p. All we required was that there *be* a threshold, and that on one side of it life would be approximately the same as usual, but on the other, we would enter a new world as yet unknown to us where gravity became quantum and all the forces and particles of nature would unify.

This is quite reasonable. General relativity reduces to Newtonian gravity in any situation where the strength of gravity isn't too overwhelming. Likewise, whatever quantum gravity turns out to be, it has to start by reaffirming and reiterating all that previous rabbis have said before, that is, it has to become indistinguishable from current theories in a first approximation and predict large new effects only in very extreme conditions: at very large energies, or for very small distances or periods. This is, after all, an observational constraint.

But then we noticed a glaring contradiction. Suppose a farmer sees a cow grazing in the fields. The cow is much larger than L_p, so the farmer can rest safe in the knowledge that his cow is uncluttered by quantum gravity hang-ups. But now Cornelia zips past in her usual mad stampede, very close to the speed of light. Cornelia conversely sees the grazing cow moving very fast with respect to her, and therefore sees her contracted in the direction of motion as predicted by special relativity. If she is moving fast enough, Cornelia may see the grazing cow contracted to a length smaller than L_p—and so conclude that she is afflicted by quantum gravitational fever, whatever that might be. Cornelia would not be altogether surprised if she saw the grazing cow start tap dancing, lap dancing, or whatever quantum gravity might cause cows to do.

But because the grazing cow is a unique entity, whatever she does should be predicted by everyone according to the same theory. Indeed, that the same theory should be used by all observers is the minimal requirement for unification. You can't have a situation in which the farmer and Cornelia need to use different theories to describe the same object; that's not just an insult to unification, it's inconsistent with the principle of relativity. If movement is indeed relative, Cornelia cannot know that she is moving and the farmer is not.

Cornelia and the farmer are again at odds, this time disagreeing over the boundary between classical and quantum gravity.

We find similar paradoxes regardless of whether we define the threshold between classical and quantum gravity with a Planck length, a Planck energy, or a Planck time. For instance, if you use the language of energies, then the problem is right in the most famous formula in physics: $E=mc^2$. As we have learnt, moving particles have higher masses, which is why we can't accelerate anything beyond the speed of light. Hence the farmer sees a stationary electron as a well-behaved particle, since its energy is much smaller than E_p; however, Cornelia will attribute a much larger energy to the electron because she sees it in motion with respect to her, and so with a larger mass. She then uses $E=mc^2$ to conclude that this larger mass translates into a larger energy; and if Cornelia moves fast enough, she may see the electron with an energy larger than E_p and therefore conclude that the electron is ruled by quantum gravitational effects. Once again, she finds a contradiction.

Lee and I discussed these paradoxes at great length for many months, starting in January 2001. We would meet in cafes in South Kensington or Holland Park to mull over the problem. The root of all the evil was clearly special relativity. All these paradoxes resulted from well-known effects such as length contraction, time dilation, or $E=mc^2$, all basic predictions of special relativity. And all denied the possibility of establishing a well-defined border, common to all observers, capable of containing new quantum gravitational effects. Quantum gravity seemed to lack a dam—its effects wanted to spill out all over the place; and the underlying reason was none other than special relativity.

The implications were unavoidable: To set up a consistent quantum gravity theory, *whatever* that might be, we first needed to abandon special relativity. We realized that many of the known inconsistencies of proposed quantum gravity theories probably also resulted from religiously assuming special relativity. Our reasoning was therefore that before doing anything clever, special relativity should be replaced by something else that rendered at least one of E_p, L_p, and t_p the same for all observers. Nothing larger than L_p should ever be contracted by motion to something smaller than L_p. Particles might appear more massive when in motion, but if their energy at rest was smaller than E_p, it should stay that way no matter how fast they were seen to move. At E_p (or L_p), all special relativistic effects should switch off, and these quantities should indeed be absolute. These were our requirements.

The difficult part was finding a new theory that could satisfy them. One thing was obvious: Whatever we did would conflict with special relativity. But as we have seen before, special relativity results from just two independent principles. One is the relativity of motion, and the other the constancy of the speed of light. Thus one solution to our puzzle could be to drop the relativity of motion; maybe at very high speeds observers would indeed become aware of their absolute motion. Some sort of ether wind would then be felt, and Cornelia would eventually realize that the farmer was at rest, while she had been in crazy flight.

This is a distinct possibility, but we decided to take the other obvious path: to preserve the relativity of motion, but admit that at very high energies the speed of light would no longer be constant. Thus VSL snuck into our arguments.

By introducing minimal modifications to special relativity, we were soon able to derive the counterpart of the Lorentz transformations in our theory. It was a lot of fun to do this. The new equations were considerably more complicated (what we call nonlinear transformations), but they were also as respectful as possible of both special and general relativity. According to our transformations, space and time would become less and less flexible the closer we got to L_p or t_p. It was as if the speed of light became larger and larger as we approached the border between classical and quantum gravity. At this border, the speed of light seemed to become infinite, and absolute space and time could be recovered, not in general, but for one specific length and time—L_p and t_p—so that everyone could agree on what belonged to classical and to quantum gravity. Our theory therefore drew an unambiguous boundary between the two realms.

Einstein's famous equation $E=mc^2$ has become so iconic that I couldn't help feeling pangs of pleasure while working out its counterpart in our theory. Hence, even though this is an appalling abuse of mathematics for a book of this type, I have to reproduce our new formula here. Bear with me and take a look at it:

$$E = \frac{mc^2}{1 + \dfrac{mc^2}{E_p}}$$

(Here c represents the approximately constant value of c we measure at low energies.) I know it's not as simple as Einstein's beauty, but if you know some basic mathematics you will soon note a remarkable property of this formula. As Cornelia flies past, she may perceive an electron resting at the farm to be as massive as she wants, if she moves fast enough. According to the usual $E=mc^2$, this means that she may see the electron with an energy larger than E_p—and so we arrive at the disconcerting conclusion that she disagrees with the farmer over whether quantum gravity is required to understand that electron.

Not so with our new formula! Even though m is not capped for Cornelia, some basic mathematics will show that according to our formula the electron's energy E can never be larger than E_p. Hence farmer and cow agree on the absence of quantum gravitational behavior in that electron.

During the cold war, every time a physicist worked out a new effect he would rush to investigate military applications. This was particularly true of U.S. physicists. Neil Turok told me he once dined at a conference with Edward Teller, and that during conversation he had mentioned to the famous physicist that he was working on magnetic monopoles. To Neil's horror, the old man at once began to work out just how much energy could be released by a magnetic monopole bomb.

Such attitudes are laughable today, of course. But to tease Lee, I computed how much energy would be released by a quantum gravitational bomb according to our formula. How much grim wealth would the phenomenally rich man withhold in our theory?

Suppose powerful accelerators managed to produce large numbers of Planck mass particles, and that somehow a bomb was made with them. According to our theory, such a bomb would release exactly *half* the energy released by a conventional nuclear weapon with the same overall mass. In other words, such an expensive quantum gravitational weapon would be precisely half as powerful as a much cheaper conventional nuclear weapon. For more massive particles (say with masses equal to twice or three times the Planck mass) the result would be even worse. I was pleased to find that even generals would probably not be dumb enough to hire Lee or me.*

AS ALL THIS EXCITING work gained shape, in the summer of 2001, guess what happened? Lee left Imperial College! Can anyone up there at Imperial recognize a bit of a pattern? Or is that too intellectually demanding?

A great controversy had broken out over the control of Lee's generous external funding. But the bad whore refused to surrender the profits, and the pimps were not amused.

As it turned out, this was only a minor irritation, as Lee had already decided to leave for Perimeter Institute (PI), in Canada. The main reason for his decision was perhaps that PI is a new research center trying to run itself in a way that is completely different from conventional science institutions. While at places such as Imperial College we constantly witness the creation of new faculties, trans-faculties, hyper-faculties and wank-faculties (as a substitute for sexual intercourse for the aging scientists who are then appointed as their directors), PI tried to flatten its structure and eliminate as many levels of management as possible. The philosophy is that since all new ideas appear to be produced by young scientists, they should also be the main steering force of science organization. As Max Perutz once said, the secret of good science is simple: no politics, no committees, no interviews—just gifted and highly motivated people. As simple as that.

---

*Unfortunately the possibility that E_p might be negative reverses this argument, as explained in our paper.

I am always suspicious of utopias, but I genuinely wish PI all the best. At the very least, they will bring bad publicity to the current corrupt scientific bureaucracies, where the recent explosion of administrative levels has ensured that administrators are now answerable only to other administrators, rather than the people they supposedly work for. Even if the PI "communist" model fails in the end, it will nonetheless show that there is something wrong with the conventional alternatives, and that someone should stop this proliferation of administrators. Personally, I would fire them all and give them a long prison sentence as a redundancy package—but you already know my thoughts on the matter.

In September 2001, I visited PI for the first time, and that's where Lee and I finished our theory. I traveled exactly one week after the September 11 attack, and I found Lee immensely disturbed by the recent events. He had just arrived from New York, where he'd visited friends living in Tribeca, and it was immediately obvious he hadn't slept the night before. I myself was unusually jetlagged, so this was a very peculiar experience.

We repaired to a bar, where we ordered beer and wine, and found ourselves on auto-repeat, making the same comments on the week's events over and over again. It was becoming so ridiculous that we eventually forced ourselves to talk physics, the only seemingly logical pursuit in a world of insanity. And, indeed, it was very soothing.

We were both so tired that we each dozed off a few times, only to wake up and find the other dozing off, too. And it was in these unlikely circumstances that the final breakthrough leading to our theory suddenly came to us! It was beautiful.*

Lee was so pleased with our results that he wanted to submit them to *Nature*. However I told him I followed a "*Nature* embargo policy":

---

*For aficionados only, it involved working in momentum space rather than real space. It turns out that the border between classical and quantum gravity is best established in terms of energy and momentum. We had been stuck on this trivial point for months.

I refuse to submit papers there until they cut off the balls of their cosmology editor. Lee laughed and suggested *Physical Review.* He also told me that *Nature,* in one of its editorials, had had the guts to accuse *Physical Review* of not publishing innovative research anymore. As a result, there was now a bit of a rift between the two journals. We laughed together at the whole thing—the misguided self-importance of all these useless people as they sing their swan songs in a world where no one gives a shit about them.

Our paper was eventually accepted by *Physical Review Letters* (not before the usual dramas), but who cares ... the important thing is that Lee and I have to this day continued to explore our theory and the explosive combination of VSL and quantum gravity.

Unlike string theory or loop quantum gravity, our work does not purport to be the final theory, indeed it assumes that we don't know it. But it does lay down, based on very simple arguments, what any consistent theory may have to assume. In the process, our more modest approach leads to concrete predictions for new observations. Could we soon test these predictions? To my mind, a bridge between quantum gravity and experiment, no matter how flimsy, is the main thing still missing. We need one desperately.

No one knows where this work might lead, but let me finish with one last story: the mystery of ultra high-energy cosmic rays. Recall that cosmic rays are particles, say protons, moving at very high speeds throughout the universe—usually the product of cataclysmic astrophysical processes such as the explosion of stars, supernovae, or even much larger detonations we still don't fully understand. Their energy range is quite disparate, but for many years it has been predicted that there should be a cutoff, a maximum energy above which no cosmic rays should be observed.

The reason is simple. In their travels throughout the universe, cosmic rays meet photons from the sea of cosmic radiation that pervades everything. These photons are very cold and their energies very low—we call them soft photons. However, if you ask what they look like from the point of view of the cosmic ray proton, well ... they look very energetic. This is a prediction of special relativity, the result of a simple calculation using the Lorentz transformations.

The faster (i.e., the more energetic) the cosmic ray, the harder and more energetic the cosmic radiation photons will appear to it. Above a certain energy, the cosmic ray protons should see the photons as having enough energy to drag stuff from their insides, producing other particles called *mesons*. In the process, the primary cosmic ray would lose some of its energy, giving it to the meson. And so any energy above the threshold for meson production would simply be shaved off.

The puzzling thing is that cosmic rays *have* been observed with energies above this cutoff, creating an anomaly that no one seems properly able to explain! But a moment of thought at once reveals that in order to work out the energy of the photons as they appear to the cosmic ray, one has to do a Lorentz transformation. The argument assumes the laws of special relativity to work out the perspective of the proton. It could be that these laws are wrong, as Lee and I suggested (as well as Amelino-Camelia and others before).

Is this the first observational mishap of special relativity, and perhaps further evidence for VSL? Could it even be our first glimpse of quantum gravity?

IT'S DIFFICULT to sum up where VSL stands, as I finish this book, because it is still well within the maelstrom of scientific inquiry. VSL is now an umbrella for many different theories, all predicting, in one way or another, that the speed of light is not constant, and that revisions to special relativity are required. Some of these theories contradict the relativity of motion—for example, the model Andy and I first proposed—but others don't. Some predict that the speed of light varies in space-time, such as my Lorentz-invariant VSL theory and Moffat's theory. In contrast, others predict that light of different colors travels at different speeds, such as the theories I developed with Stephon and Lee. It is also possible to overlay some of these theories and have both space-time *and* color variations in c. Some of these theories were designed as cosmological models, others as black hole theories, and yet others as solutions to quantum gravity.

And this is only a small sample. In the Web archives used by physicists there has now accumulated a large literature on the subject. Indeed, I have recently been asked to write a long review of all the VSL ideas that have been proposed to date. This is, I hope, a sign of maturity—not senility.

The reason for this diversity is that we simply don't know which, if any, of these theories is correct. There are also hundreds of models of inflation, and until convincing proof of inflation is found, the situation is unlikely to change. But VSL is different because, unlike inflation, it has a lot to say about physics that can be tested here and now. It's not just some primordial fart of the baby universe; VSL should reveal itself in subtle effects directly accessible to experimentalists. The most direct example is Webb's changing alpha observations. But the current acceleration of the universe could be another tell-tale signature of the class of VSL theories that predict variations of c in space-time.

So far, all these connections with experiment are postdictions, but my current VSL work is mostly devoted to predictions. There is no more effective way to shut skeptics' mouths than to predict a new effect and then verify it by experiment. In this vein, John Barrow, my student Håvard Sandvik, and I have done a lot of work showing that the value of alpha, as measured from spectral lines formed in compact stars or black hole accretion disks, should also be different. An observation of this effect would be a quite spectacular vindication of VSL. We have also found that certain VSL competitors that can explain Webb's results predict small violations of Galileo's principle stating that all objects fall in the same way. A satellite experiment (called STEP) could soon disprove these competing theories, which attribute the variations in alpha to a variable electron charge, rather than to c. We eagerly await these new observations.

In contrast, ultra high-energy cosmic rays (and similar anomalies found by astronomers) have more to say about theories predicting color variations in c. These theories also imply other new phenomena, such as the correction to $E=mc^2$ I told you about. And every time I find a new prediction, I rush to locate the experimental physicists who might be able to measure it. More often than not they tell me I'm crazy, and that with current resources there is no way we can measure such tiny effects. But I am more optimistic than they are and have always felt that experimental physicists are cleverer than they think. Perhaps proof that VSL is right might not be too far off.

And what if it is wrong? It is amusing that some of my colleagues—to be fair, a minority—are *desperate* to see VSL fall on its face. These are the people who've never had the balls to try to find something truly new themselves. It's a sad fact that some scientists never stray far from the path of what is already known, be it in string theory, inflation-based cosmology, or cosmic radiation theory and experiment. Clearly something as wild as VSL is an affront to their self-respect; so they *need* to see it fail. But they miss the point. If it fails, I'll just try again with something even more radical, because it's the process of losing oneself in the jungle that makes science worth doing.

Needless to say, if VSL turns out to be correct, these people will at once sharply deny all their previous slurs and start working on it. They are bandwagon passengers, those who play safe and lead an easy life, generously rewarded by funding agencies and the scientific establishment. John Barrow once remarked that any new idea goes through three stages in the eyes of the scientific community. Stage 1: It's a pile of shit and we don't want to hear about it. Stage 2: It's not wrong but it certainly has no relevance whatsoever. Stage 3: It's the greatest discovery ever made and we found it first. Undoubtedly if VSL is correct, there will be no shortage of current detractors who will twist history to claim priority.

Just as surely, I'll be on yet another intellectual adventure.

EPILOGUE: FASTER THAN LIGHT

AS THIS BOOK goes to press, no one knows whether VSL is right or not, and if it is correct, in which particular incarnation. Neither do we know what its most immediate implications might be: cosmology, black holes, astrophysics, quantum gravity? The current observational evidence for VSL—John Webb and collaborators' findings, the supernovae results, and ultra high-energy cosmic rays—remains controversial. But even if these observations turn out to be an illusion contrived by experimental errors, some VSL theories will still survive; the field will just be less exciting. There may also be other observations lurking just around the corner, waiting to vindicate or refute the theory. It's all up in the air.

I am often asked whether this state of affairs is unnerving and whether it will be humiliating for me if VSL is disproved. My answer is invariably that there is no humiliation in seeing your theory ruled out. That's part of science. The important thing is to *try* new ideas, and regardless of what happens to VSL that's what I've done. I've struggled to expand the frontiers of knowledge by jumping into that grey area where ideas are not yet right or wrong, but are mere shadows of "possibilities." I've thrown myself into the darkness of speculations, and thus participated in the Big Detective Story so vividly described in *The Evolution of Physics,* that bright present my dad gave me so many years ago. I will never regret what I did.

Another reason for having no regrets are the extraordinary people VSL has revealed to me. All the characters in this book have become close friends, and I keep in touch with all of them. This alone has made it worthwhile.

I never worked with Andy again, but he has remained my guru. I always call him before going overboard with science policy matters, but funnily enough in recent times his advice has become very anarchist indeed. Andy has kept an outside interest in VSL, but has moved on to more mainstream ideas. He is currently at UC Davis, successfully setting up a new cosmology group, and he and his family seem very happy with their new life. Still, I have occasionally caught Andy grumbling that his current students are not like those he had in the good old days at Imperial College.

John Barrow has moved to Cambridge, where he lives in palatial quarters and holds a professorship at the university. He still writes a book and at least ten scientific articles each year. We have continued to work together, on and off, on all sorts of "varying constant" theories, including VSL. He comes down to London regularly, and we meet at the noble headquarters of the Royal Astronomical Society, which is a sort of English gentlemen's club. We brainstorm, gossip, and giggle while sipping wine, and then write a new paper together. What a British way to do science. . . .

Kim never returned—or expressed any desire to return—to the falsely glittering world of academia. Although I was greatly upset when she left, I have to admit that there is far more to life than the fucking secrets of the universe. Kim currently teaches mathematics at a girl's school in South London, the sort of worthwhile job hundreds of useless education politicians should be doing, but can't. She still puts up with me; no one else would.

Stephon is still at Imperial, but he is about to move back to the United States as a postdoctoral fellow at Stanford. He's more than ever a force of nature, the source of an endless stream of new ideas in cosmology and string theory. We haven't been to the Globe for a while, but the other day Stephon sighted His Highness the Eagle driving a shiny new racing BMW. Stephon has a long-term project to launch a new research institute in Trinidad: CIAS (Carribean Institute of Advanced Study). It is in this sort of initiative that the future of science lies.

But the person I'm closest to now is Lee, a result of the developments I recounted toward the end of this book. Indeed, the story is still unfolding as we keep working on our rendition of quantum gravity, finding more and more interesting results. We meet in London or at PI, which has gone from strength to strength and is now one of the most bubbling scientific centers in the world. PI is currently housed in a former restaurant—complete with a bar and pool table—which may have something to do with this.

Another regular visitor there is John Moffat, now retired from Toronto University but still as prolific as ever. Some of the most exciting recent VSL ideas originated with him and his coworker Michael Clayton. I keep nagging him to write his memoirs; he has lived through so much, happy and sad in equal measures. He is also our last link with the golden generation of Einstein, Dirac, Bohr, and Pauli, and believe me, the stories I have included in this book are but a fraction of what he has to offer. Still, I have recently realized why he hasn't tried to record them: Physics is only a small part of his story. When it comes down to it, we are all larger than life.

With such a sparkling cast, I am aware that I've told you more private details than is usual in a book of this kind. I trust this has conveyed the feeling that doing science is not only fun but also a remarkable human experience that brings people together. In this respect, the VSL story may be lively, but is far from extreme, as I realized when I unearthed the story behind my childhood favorite—*The Evolution of Physics* by Einstein and Infeld—while doing research for this book.

Leopold Infeld was a Polish scientist who worked with Einstein on various important scientific problems in the 1930s. Einstein began to act as his mentor, and as it became obvious that a German invasion of Poland was in the cards, Einstein realized what lay ahead for Infeld should he remain there. Naturally, Einstein took it upon himself to save his friend. But by the late 1930s, he had supported the immigration of so many Jewish families that his affidavits were essentially worthless in the eyes of the U.S. authorities; in particular, they ignored his pleas on behalf of Infeld. Einstein tried to find him a professorship at a U.S. university, but times were tough and that failed, too. As tensions mounted in Europe, Infeld's prospects looked very grim indeed.

Out of desperation, Einstein hit upon the idea of writing a popular science book in partnership with Infeld. This was nothing less than *The Evolution of Physics*—the book that many years later, with its unique beauty, would seduce me into becoming a physicist—written in great haste in just a couple of months, to become a huge sensation, making Infeld suddenly desirable to the U.S. authorities. Without this success, Infeld would most likely have gone up in smoke in some Nazi hell.

The VSL story may not be quite so dramatic, but I hope I have shown that science is above all a rewarding human experience, perhaps the purest one on offer in a world too often less than perfect. In addition, I hope I have brought visibility to what actually happens while new science is being done. Believe me, it's nothing like the rational chains of logic that historians of science like to credit us with. If you look at the "farmers," such chains may seem a fair account of reality, but if you consider the "pioneers," the story is very different. It's all about fumbling in the dark, trying and trying and more often than not failing spectacularly, but always being madly in love with your quest, excited beyond measure by what you do.

As I write these final words under the blue skies of West Africa, I think of an old woman I met yesterday in a remote village. She is the great-grandmother of a friend of mine, who is in his late twenties.

No one knows her age exactly; in her day, no one bothered counting years or trying to measure the space between cradle and coffin. She looked wise and beautiful as she talked in her rich, soft voice, mixing the musical sounds of Mandinka with expressive grunts and pauses, while gazing around with strangely disturbing eyes (I would later learn that she was blind).

As is true with many old people, she was fond of recalling her youth, in a faraway time when no one had ever seen a white person around that village, even though in the eyes of the British the Gambia was their colony—a good example of a power delusion. She claimed that life was better and people were happier in those days. When I asked why, she replied, "Because there was more rice."

Traveling around that village reminded me further of the fifteenth-century Portuguese sailors who traded cheap mirrors for gold from local African tribes. The sailors thought they were cheating the Africans, but think about it. The value of gold is purely conventional, an unwritten agreement originally peculiar to European and Asian cultures. As far as I know, no one eats gold. What the African elders thought of this roaring trade is not on record, but it's conceivable that they felt they were cheating the sailors. They were giving them useless pieces of rock in return for fine contraptions that enabled them to see their reflections.

Between scientists and the establishment there is often a similar "cultural" trompe l'oeil. They think they own us; we think we have it all and that they're just a bunch of squares. They may have the power, the easy success, and the impression that they're in control, but we think they're badly deluded. It's we—who love the unknown beyond any trends, politics, or party lines—who have the last glorious laugh. We love our work beyond anything words can describe. We have all the fun in the universe.

ACKNOWLEDGMENTS

THIS BOOK would not have been possible without the help of Kim Baskerville, Amanda Cook, and Susan Rabiner, who taught me how to read and write in this beautiful language. I thank them for never losing their patience, which must have been very difficult indeed.

Then, of course, this book would not exist without its characters. I thank my comrades in arms Andy Albrecht, John Barrow, John Moffat, Stephon Alexander, and Lee Smolin. Before being great scientists they are great people, and I thank them for their warm friendship.

But to be honest I would never have thought of popularizing VSL were it not for the interest it excited in the media. In this respect I am grateful to many people in various countries, but am above all indebted to David Sington, the producer/director of the documentary "Einstein's Biggest Blunder," who showed me the way to this book.

While writing the book I gave a large number of talks in schools. Little did these innocent pupils know that I was using them as guinea pigs to test my arguments. I am very grateful to many a sharp young mind for showing me when I was being dumb.

Also, for reading drafts of this book and providing valuable comments, thanks are due to Kim, David, Andy, the two Johns, Stephon, and Lee.

The paperback edition further benefited from helpful remarks by M. French, J. Gribbin, P. Ivo Teixeira, J. Michael West, and L. Robinson.

Very little of the book was written in London, since I tried to shield my scientific research from its development. As a result my scribblings took place all over the shop, exacerbating my nomadic tendencies. I would like to thank the hospitality of all those who welcomed me everywhere, in particular Gianna Celli of the Rockefeller Center in Bellagio.

Finally I would like to dedicate this book to my father, Custodio Magueijo, who bought me all those crazy books when I was a kid. Although there are many remarkable debts listed above, I still think this is the greatest of all.

CREDITS

I WOULD LIKE TO THANK Paul Thomas for the cartoons in Chapter 2; Meilin Sancho for the photos of Andy Albrecht and Stephon Alexander; Pembrey Studio, Cambridge, for John Barrow's photo; Patricia Moffat for her husband's photo; Dina Graser for Lee Smolin's photo; the Astrophysical Research Consortium and the SDSS Collaboration for the NGC6070 galaxy picture; O. Lopez-Cruz, I. Sheldon, and the NOAO/AURA/NSF for the optical picture of the Coma cluster; the ROSAT Science Data Center and the Max-Planck-Institut fur Extraterrestrische Physik for the X-ray picture of the same cluster; and the NASA Goddard Space Flight Center and the COBE Science Working group for the DMR map.

INDEX

Absolute motion, 250
Action principles, 190–191
African colonies, 263
Aging process
  black holes and, 228
  time dilation effect and, 33–34
Albrecht, Andy
  author's relationship with, 136,
    138–140, 178–181, 202, 260
  inflationary theories and, 8, 124,
    139
  life at Imperial College, 148–149
  photograph of, 137
  string theory and, 239
  VSL theory and, 150–152,
    166–167, 202–203, 220
Alexander, Stephon, 240–241, 242,
  260–261
Alpha, 198–199, 257
Alpha-Centauri, 33
Amelino-Camelia, Giovanni, 246
Antimatter, 110
Argumentative tradition, 2
Aristotle, 49
Aspen conference, 165–175
Atomized space-time, 244
Australia
  aboriginal songlines of, 22

author's trip to, 192–197
VSL discoveries from, 195–196

Barrow, John
  author's relationship with,
    191–192, 194–195, 220, 260
  photograph of, 192
  remark about new scientific ideas,
    258
  scientific journals and, 197, 206,
    217
  VSL theory and, 192, 199, 220,
    257
Bianchi identities, 157
Big Bang theory
  bouncing universe and, 106, 107
  cosmological constant and,
    99–105, 231
  flatness problem and, 83, 94–99,
    119–122
  horizon problem and, 80–83,
    116–119
  Hot Big Bang model and, 104
  Hubble's law and, 78
  inflationary universe and, 115–116,
    122–123
  particle physics and, 109–111
  problems with, 3

riddles of, 79–80, 105, 106–108,
    132
Big Crunch, 66, 73, 106
Big Issue magazine, 135
Black holes, 226
    escape velocity for, 226–227
    horizon of, 227
    time and, 227–228
Bohr, Niels, 212, 237
Bonaparte, Napoleon, 31n
Bond, Dick, 210
Born, Max, 43
Bouncing universe, 106, 107, 171
Brandenberger, Robert, 241
Brane-world cosmology, 242
Bureaucracy, 142, 148

Cambridge University, 129–130, 149,
    222
Candide (Voltaire), 188
Churchill, Winston, 31n
Clayton, Michael, 216n, 261
Clockwork universe, 21
Closed universe, 91, 93, 158
Color-dependent speed of light,
    244–245, 246–247, 256
Coma cluster, 100, 101
Conservation of energy, 156–158
Contact forces, 46–47
Cornish, Neil, 210
Cosmic density, 95–96
Cosmic dust, 103
Cosmic expansion, 87–89
Cosmic radiation, 101–102, 103–104,
    172
Cosmic rays, 110, 255
Cosmic strings, 228–229
Cosmological constant, 67, 99
    cosmic expansion and, 230–231
    flatness problem and, 119–121
    inflation and, 123
    vacuum energy and, 104–105, 162

    varying speed of light and,
        161–162
Cosmological fluid, 65, 76, 89
Cosmological perturbation theory,
    171–173
Cosmology, 14–15
    Big Bang theory and, 3, 80
    Hubble's discoveries and, 76
    inflation theory and, 3–4
    particle physics and, 111–114
    quantum gravity and, 245–246
    science of, 15, 96
    VSL theory and, 152–156
Critical density, 95–96, 177
Curved space-time, 51–53, 234

Dark matter, 101, 103
Davies, Paul, 194
"Death of Scientific Journals, The"
    (Magueijo), 217–219
Deflection angle, 63
Density, critical, 95–96, 177
Density contrast, 172
"Diatribe of Dr. Akakia, The"
    (Voltaire),
    189
Dicke, Robert, 96, 121
Differential geometry, 42, 52–53
Dirac, Paul, 110, 130
Drummond, Ian, 216n
Durrer, Ruth, 138–139

$E=mc^2$ equation, 35–37, 158, 249,
    251, 257
Eccentricity, 56
Eclipses, solar, 61–64
Einstein, Albert
    bovine dream of, 15–21
    contemporary physics and,
        236–237
    early adult life of, 29–32

Evolution of Physics book by, 13, 261–262

Friedmann's work and, 85–86

general relativity theory of, 41–43, 62–64, 71–72

immigration to the U.S. by, 170

Meaning of Relativity book by, 43–44

Moffat's work and, 213–214

special relativity theory of, 6, 26, 31–38, 70

static universe theory of, 73

time dilation effect of, 228, 229

unified theory of, 10, 213, 233

Electric forces, 47

Electromagnetism, 47, 234

Elliptical orbits, 56

Energy
   conservation of, 156–158
   mass and, 36–38
   speed of light and, 35–36, 247

Escape velocity, 66, 226–227

Eternal universe, 231

Ether theory, 68–70

Euclidean space, 89

Evolution of Physics, The (Einstein & Infeld), 13, 43, 70n, 260, 261–262

Expanding universe
   cosmological constant and, 99–105
   flatness problem and, 96–97, 119–122
   gravity and, 93–94
   horizon problem and, 80–83, 116–119
   inflation and, 115–116, 118
   models of, 87–93
   Planck time and, 97–98

Expansion factor, 88

Extra Dimensions, 144, 239, 242

Fado singing, 155

Faraday award, 220

Fast-light, 133–134

Fast-tracks, 228

Feynman, Richard, 168

Financial issues, 149–150

Fine structure constant, 198–199

Five-dimensional space-time, 144

Flatness problem, 83
   description of, 94–99
   homogeneity problem and, 177
   inflationary solution to, 119–122
   VSL theory and, 176–177

Flat universe, 91, 93–95, 158

Forces
   contact, 46–47
   electric, 47

Four-dimensional space-time, 34–35, 144

Fractal people, 137–138

Frequency-dependent speed of light, 244–245

Friedmann, Alexander, 15, 83–86, 105n

Fundamental strings, 238

Galaxies
   clusters of, 100
   discovery of, 75–76
   motion of, 77–78

Galileo Galilei, 43, 48, 49, 257

Gamow, George, 35

General theory of relativity
   discovery of, 38–39
   Friedmann's study of, 84–85
   gravity and, 41–43, 62–64, 216
   massless particles and, 216
   vacuum and, 71–72
   See also Special theory of relativity

Geodesics, 51–52

Geometry
   differential, 42, 52–53
   noncommutative, 244

Global positioning system (GPS), 22

Goa, India, 153–155
Grant proposals, 141–142
Gravitational lens effect, 61–64
Gravitational mass, 50
Gravitons, 215–216, 234
Gravity
    attractive nature of, 66, 71
    curved space-time and, 51–53, 234
    expanding universe and, 93–94
    Galileo's experiments with, 49
    general relativity theory and,
        41–43, 62–64
    light bent by, 60–64
    massless particles of, 215–216
    Newtonian theory of, 46, 47–48,
        50, 55–60
    nonsymmetric metric theory of,
        235
    quantum, 10, 234–238, 245–252
    special relativity theory and, 41,
        47–48
    speed of light and, 47–48, 251
    supercooled matter and, 114
    universal prevalence of, 67–68
Guth, Alan
    Big Bang riddles and, 108
    career of, 111, 125
    collaboration with Henry Tye,
        112–114
    influence of Robert Dicke on, 96
    monopole problem and, 112–114,
        124, 228
    theory of inflation by, 3, 111–115,
        121, 123

Hindmarsh, Mark, 107
Hitler, Adolf, 243
Hoffmann, Banesh, 42
Homogeneity problem
    flatness problem and, 177
    horizon problem and, 170–171,
        173

Homogeneous universe, 82, 89, 120,
    137, 173
Horizon problem
    description of, 80–83
    homogeneity problem and,
        170–171, 173
    inflationary solution to, 116–119
Hot Big Bang model, 104, 245
Hoyle, Fred, 212
Hubble, Edwin, 15, 73–78, 87
Hubble's law, 77–78, 88
Humason, Milton, 74
Hypersphere, 92

Imperial College
    academic wages at, 148–149
    administrative duties at, 150
    author's tenure at, 219–220
    development of VLS theory at,
        159
    leadership problems at, 203–204,
        252
    scientific environment at, 203
Inertia
    law of, 200, 201
    Newtonian gravity and, 50
Inertial mass, 50
Infeld, Leopold, 13, 261–262
Inflation
    Big Bang models and, 122–123
    cosmological problems and,
        3–4
    discovery of, 115–116, 123
    finding alternatives to, 4–5,
        132–133, 139
    fixing flaws in, 123–124
    flatness problem and, 119–122
    horizon problem and, 116–119
    Lambda problem and, 120–121,
        123
    quantum gravity and, 237
Inflationary Universe, The (Guth), 124

Inflaton field, 124, 132
Inquisition, 190

Journals, scientific, 217–218

Kaluza-Klein theories, 144–147
Kepler, Johannes, 56
Kepler's first law, 56
Kibble, Tom, 167, 179
Kim (author's girlfriend), 142–143,
    175, 178, 193, 221–223
Kinetic energy, 36
Kiritsis, Ellias, 241, 242

Lambda, 67, 99
    cosmic expansion and, 230–231
    flatness problem and, 119–121
    inflation and, 123
    vacuum energy and, 104–105
    VSL theory and, 161–162
Law of inertia, 200, 201
Leaning tower of Pisa, 48–49
Least action principle, 187–188
Leibnitz, 188
Length contraction effect, 27, 29, 249
Le Verrier, Urbain-Jean-Joseph, 58
Levin, Janna, 210, 217
Light
    characteristics of, 214–215
    effect of gravity on, 60–64
    See also Speed of light
Linde, Andrei, 245
Local speed limit, 225
London, Jack, 30
Look-back time, 81
Loop quantum gravity, 236
Lorentz invariance, 214, 215,
    224–225
Lorentz symmetry, 190–191,
    200
Lorentz transformations, 190,
    255

Magellanic Clouds, 76
Magnetic monopoles, 109–110,
    112–114, 124, 228, 252
Magnetism, 234
Martin Eden (London), 30
Mass
    energy and, 36–38
    gravitational, 50
    inertial, 50
    photons and, 215
    speed of light and, 36
Massless particles, 215–216
Matter
    dark, 101, 103
    origin of, 159
    supercooled, 114
    types of, 101
Maupertuis, Pierre de, 187–190
Mavromatos, Nikos, 246–247
Meaning of Relativity (Einstein), 43–44,
    200
Membranes, 241–242
Mercury (planet), 58–59
Mesons, 255
Michelson, Albert, 5
"Micromegas" (Voltaire), 190
Microwaves, 101
Milky Way, 76
Minkowski space-time, 35, 107–108
Moffat, John, 209–218, 237, 261
Momentum space, 254n
Monopole problem, 109–110,
    112–114, 124, 228
Morley, Edward, 5
Motion
    absolute, 250
    energy and, 36
    photons and, 215
    relativity of, 250, 256
    space and, 27–29
    time and, 28–29
Mount Stromlo observatory, 194

Mount Wilson observatory, 74
M-theory, 238, 239–240, 241
Muons, 29

Nature (journal), 183–184, 185n, 254
Neptune, 58
Neutrinos, 102n, 215
Newton, Isaac, 15, 46, 56, 186–187
Newtonian theory of gravity, 46
  inertial mass and, 50
  planetary orbits and, 55–60
  special relativity theory and, 47–48
Noncommutative geometry, 244
Nonsymmetric metric theory of
  gravity, 235
Notting Hill, 242–243

Omega, 94–95, 96, 98, 121–122
Open universe, 91, 92, 120, 158
Orbits, planetary, 55–60

Particle physics
  Big Bang theory and, 109–111
  cosmology and, 111–114
  phase transitions in, 112
Peer review process, 183–186,
  218–219
Perimeter Institute (PI), 253
Perutz, Max, 253
Phase transitions, 112, 160
Phoenix universe, 106
Photons, 215, 244, 255
Physical Review D (PRD), 185, 192,
  197, 202
Physical Review Letters, 254
Physics
  circular arguments in, 200
  universal mysteries and, 14
Pi, 198–199
Pietronero, Luciano, 138
Pine, Courtney, 141
Plagiarism, 219

Planck energy, 247
Planck epoch, 97–98, 160, 161, 235
Planck length, 234, 238, 247
Planck time, 97–98, 160, 234, 247
Planetary orbits, 55–60
Political correctness, 222
Polyakoff, Serge, 212
Portuguese colonialism, 153
Postdiction, 55
Poverty, 149
Prediction, 55
Princeton University, 136
Principia (Newton), 56, 186
Proportionality constant, 157
Pseudosphere, 91–93

Quanta, 233–234
Quantum gravity, 10, 234–238
  cosmology and, 245–246
  exploring the threshold of,
    247–248
  special theory of relativity and,
    249–250, 255
  VSL theory and, 234–235,
    237–238, 246–252, 254
Quantum theory, 146, 233–234

Refereeing process, 183–186,
  218–219
Relativity theory. *See* Theory of
  relativity
Repulsive vacuum, 71
"Rich man" analogy, 38
Riddles of the Big Bang, 79–80, 105,
  106–108, 132
Royal Astronomical Society, 260
Royal Society, 136, 220
Russian scientists, 83

Salam, Abdus, 130, 212, 213
Sandvik, Håvard, 257
Sarkar, Subir, 247

Scale factor, 88, 90–91, 93
Schröedinger, Erwin, 212
Sciama, Dennis, 212
Science
    mysteries of, 13
    predictions in, 55
Scientific journals, 217–218
Second law of thermodynamics,
    106n
Sexism, 222–223
Sleep habits, 31n
Smolin, Lee, 123, 124, 246–247,
    252–254, 261
Songlines, 22
Sound, speed of, 18–19
Space
    curvature of, 53
    extra dimension of, 144, 145
    movement of objects in, 27–29
    units for measuring, 21–22
Space-time, 6
    atomized, 244
    curved, 51–53, 91, 234
    Kaluza-Klein model of, 144–147
    Minkowski, 35, 107–108
Space travel
    speed of light and, 33–34
    time dilation effect and, 33, 229
    VSL cosmic strings and, 229
Special theory of relativity, 6
    discovery of, 31–32
    $E=mc^2$ equation and, 35–38
    ether theory and, 69, 70
    implications of, 32–38
    massless particles and, 215
    quantum gravity and, 249–250,
        255
    revisions to, 256
    VSL contradiction of, 190, 256
    See also General theory of
        relativity
Speculation, 1

Speed of light
    black holes and, 226–227
    color-dependent, 244–245,
        246–247, 256
    constancy of, 5–6, 26, 32, 200,
        201, 250
    cosmic speed limit as, 32–33, 35,
        81
    cosmic strings and, 228–229
    fast-light and, 133–134
    gravity and, 47–48, 251
    mass and, 36
    massless particles and, 215–216
    rate of change of, 153
    space travel and, 33–34
    speed of sound vs., 18–19
    time and, 27, 197–198, 224–225
    varying of, 6–7, 42, 156, 231
    See also Varying speed of light
        (VSL) theory
Spherical universe, 91, 93
Static universe, 67, 73
Steinhardt, Paul, 124, 139
St. John's College, 130
Strings, 228–229
String theory, 236, 238–241
Strong force, 234
Sun
    deflection of light rays by, 60–61
    eclipses of, 61–64
Supercooled universe, 113–114, 115
Swansea, South Wales, 175

Tautologies, 201, 225
Teaching Quality Assessment (TQA),
    204–205
Teller, Edward, 252
Tension, 72
Thatcher, Margaret, 31n
Theory of everything, 233
Theory of relativity
    cosmological discovery and, 15

E=mc² equation and, 35–38
gravity and, 41–42
implications of, 32–38
See also General theory of
relativity; Special theory of
relativity
Thermodynamics, second law of,
106n
Thomas, Dylan, 175
3-brane cosmology, 242
Threshold of quantum gravity, 247
Time
belief in constancy of, 21
black holes and, 227–228
cosmic strings and, 229
distance and, 80–81
movement of objects and, 28–29
relativity of, 18, 27
speed of light and, 27, 197–198,
224–225
units for measuring, 23, 224
See also Space-time
Time dilation effect, 228, 229, 249
Topological defects, 4
Trinity College, 131, 213
Turok, Neil, 106–108, 137, 141, 179,
204, 252
Twin paradox effect, 229
Tye, Henry, 112–114

Ultra high-energy cosmic rays, 255,
257
Unified theory, 10, 213, 233–234,
248
Units of measure, 21–22, 224
Universe
age of, 81
bouncing, 106, 107
clockwork, 21
closed, 91, 93, 158
eternal, 231

expanding, 87–89
flat, 91, 93–95, 158
homogeneous, 82, 89, 120, 137,
170–171
inflationary, 115–116
open, 91, 92, 120, 158
restless, 65–66
static, 67, 73
supercooled, 113–114, 115
Uranus, 58

Vacuum energy, 70, 71–72, 104–105,
162, 231
Varying constant theories, 191
Varying speed of light (VSL) theory
action principle formulation of,
190–191
conservation of energy and,
156–158
constructive criticism of,
197–198
cosmological constant and,
161–162, 231
cosmological perturbation theory
and, 171–173
development of, 135–140,
150–153, 156–163
Einstein's proposal of, 42
flatness problem and, 176–177
homogeneity problem and,
170–171, 173, 176
initial responses to, 6–7
Kaluza-Klein theories and,
145–147
Lorentz invariant version of,
224–225
models of, 223
Moffat's work on, 209–212, 214
M-theory and, 241–242
observational evidence of, 196,
198–199

original ideas about, 133–135
peer review process for, 183–186
phase transition and, 160–161
publication of, 206–207, 254
quantum gravity and, 234–235,
    237–238, 246–252, 254
special relativity theory and, 190,
    256
Visible light, 214
Volkas, Ray, 194
Voltaire, 188–190
Vulcanus, 59

Weak force, 234
Webb, John, 195, 198–199, 206, 256,
    259
World-line, 35
World Wide Web
    copyright issues and, 219
    scientific papers on, 218–219
    VSL theories on, 256

X rays, 100, 101

Zeldovich, Yakov, 105–106, 171

FOR THE BEST IN PAPERBACKS, LOOK FOR THE

In every corner of the world, on every subject under the sun, Penguin represents quality and variety—the very best in publishing today.

For complete information about books available from Penguin—including Penguin Classics, Penguin Compass, and Puffins—and how to order them, write to us at the appropriate address below. Please note that for copyright reasons the selection of books varies from country to country.

In the United States: Please write to *Penguin Group (USA), P.O. Box 12289 Dept. B, Newark, New Jersey 07101-5289* or call *1-800-788-6262.*

In the United Kingdom: Please write to *Dept. EP, Penguin Books Ltd, Bath Road, Harmondsworth, West Drayton, Middlesex UB7 0DA.*

In Canada: Please write to *Penguin Books Canada Ltd, 10 Alcorn Avenue, Suite 300, Toronto, Ontario M4V 3B2.*

In Australia: Please write to *Penguin Books Australia Ltd, P.O. Box 257, Ringwood, Victoria 3134.*

In New Zealand: Please write to *Penguin Books (NZ) Ltd, Private Bag 102902, North Shore Mail Centre, Auckland 10.*

In India: Please write to *Penguin Books India Pvt Ltd, 11 Panchsheel Shopping Centre, Panchsheel Park, New Delhi 110 017.*

In the Netherlands: Please write to *Penguin Books Netherlands bv, Postbus 3507, NL-1001 AH Amsterdam.*

In Germany: Please write to *Penguin Books Deutschland GmbH, Metzlerstrasse 26, 60594 Frankfurt am Main.*

In Spain: Please write to *Penguin Books S. A., Bravo Murillo 19, 1° B, 28015 Madrid.*

In Italy: Please write to *Penguin Italia s.r.l., Via Benedetto Croce 2, 20094 Corsico, Milano.*

In France: Please write to *Penguin France, Le Carré Wilson, 62 rue Benjamin Baillaud, 31500 Toulouse.*

In Japan: Please write to *Penguin Books Japan Ltd, Kaneko Building, 2-3-25 Koraku, Bunkyo-Ku, Tokyo 112.*

In South Africa: Please write to *Penguin Books South Africa (Pty) Ltd, Private Bag X14, Parkview, 2122 Johannesburg.*